S. Robbert Gradstein, Jürgen Homeier and Dirk Gansert (Eds.)

The Tropical Mountain Forest

erschienen als Band 2 in der Reihe „Biodiversity and Ecology Series" des
Göttingen Centre for Biodiversity and Ecology
Universitätsverlag Göttingen 2008

S. Robbert Gradstein, Jürgen
Homeier and Dirk Gansert (Eds.)

The Tropical
Mountain Forest

Patterns and Processes in a
Biodiversity Hotspot

Biodiversity and Ecology Series:
Volume 2

Universitätsverlag Göttingen
2008

North American Edition
The University of Akron Press

First University of Akron Press edition, 2010.

ISBN: 978-1-931968-79-9
Distributed exclusively by The University of Akron Press in the United States and Canada.
The paper used in this publication meets the minimum requirements of American National
Standard for Information Sciences—Permanence of Paper for Printed Library Materials,
ANSI Z39.48–1984. ∞
Manufactured in the United States of America.

First published 2008 in Germany by Universitätsverlag Göttingen

Contents

Contents

Preface

Tropical mountain forests (or "montane forests"; the two terms are interchangeable although the latter more specifically refers to the altitudinal vegetation belt) are very rich in species and are generally considered as hotspots of biodiversity. They are also of great ecological importance as sources of water and other ecosystem services for millions of people living in the tropics. However, these valuable forest ecosystems are now increasingly being fragmented, reduced and disturbed by human interventions. Concern about the future of the tropical mountain forests has triggered an increasing amount of research on the exceptionally rich biodiversity and ecological complexity of these forests in recent years.

This book originated from a lecture series on the tropical mountain forest organized by the Göttingen Centre of Biodiversity and Ecology and held at the University of Göttingen, Germany, in the summer of 2007. The purpose of the lecture series was to present a synthesis of current ecological research in Germany on the tropical mountain forest, from an interdisciplinary perspective.

The subjects presented include a large variety of topics including climate, quaternary history, species richness and endemism, impact of forest disturbance on biodiversity, mycorrhizal diversity, soil fauna, vegetation dynamics, carbon allocation and productivity, forest hydrology, soil dynamics, indigenous land use and sustainable management of tropical mountain forests. The final chapter summarizes current understanding of the incidence of tropical mountain forest hotspots from an ecosystem perspective. All contributions are based on recent empirical research, with a special focus on the Andes of Ecuador which harbour one of the richest and most endangered mountain forest resources on earth and are the location of major ongoing investigations of Research Units 402 and 816 (www.tropicalmountainforest.org) of the German Research Foundation (DFG).

We are much indebted to the authors for their valuable contributions. We also thank the DFG Research Unit 816 and its speaker, Prof. Dr. J. Bendix, for financial support towards the publication of this book.

Göttingen, February 2008

Stephan Robbert Gradstein
Jürgen Homeier
Dirk Gansert

Biodiversity and Ecology Series (2008) 2: 7-24
The Tropical Mountain Forest – Patterns and Processes in a Biodiversity Hotspot
edited by S.R. Gradstein, J. Homeier and D. Gansert
Göttingen Centre for Biodiversity and Ecology

Tropical mountain forests – distribution and general features

Michael Richter

Institute of Geography, University of Erlangen, Kochstr. 4/4, 91054 Erlangen, Germany,
mrichter@geographie.uni-erlangen.de

Abstract. Tropical mountain forests vary considerably in continental occurrence and vertical extension. They are most widespread in South America and in (semi-)humid mountain areas. In terms of phytogeography, temperate floristic elements become increasingly important towards mountains of the outer tropics. Species richness is much higher in humid tropical mountains than in dry ones. In terms of biodiversity, tropical mountain forests are one of the world's main hot spots; this high biodiversity is triggered at different scales. At the world scale, the great length of effective evolutionary time, the constant input of "accidentals" and the type of seasonal variability in tropical mountain forests are of importance. At the landscape scale, the input of taxa from various directions along mountain chains, habitat fragmentation and geological history contribute to diversification. At the mountain scale, the great variety of climatic characteristics, orographic heterogeneity and geological and edaphic conditions are important. At the belt scale, disturbance regimes and stress factors, orographic atributes and local altitudinal climate gradients are crucial factors. Finally, at the patch scale alpha-diversity is triggered by micro-habitat differentiation as exemplified by epiphytic niche variety, and micro-relief and succession patterns. Due to the species-rich tree communities, discrete vegetation borders are barely identifiable.

Introduction

Nearly all of the world´s mountain chains harbour mountain forests. Only a few extremely dry or cold mountainous regions such as the central and southern part of the Cordillera de Atacama, the western and central Tibet highlands, and Svalbard lack in tree growth. The forested mountains differ considerably in vegetation and three main groups may be distinguished: 1) Extratropical mountains, with mixed and pure coniferous forests in the northern and deciduous forests in the southern hemisphere; 2) Subtropical mountains, with evergreen as well as deciduous broadleaved forests in the more humid and coniferous woodlands in the drier ecozones; and 3) Tropical mountains, with evergreen and semi-deciduous forests.

The latter mountains possess various climate and soil characteristics differing considerably from extra- and subtropical areas. A main feature is the absence of seasons with long periods of snow and frost at mid-altitudes. From a thermal point of view, this is due to the "tropical daytime climate" (Troll & Paffen 1964). In the tropics, seasonality is primarily based on discontinuous precipitation regimes. However, dry "summers" are more pronounced in the lowlands than in the uplands. Thus, most of the tropical mountain forests profit from higher rainfall inputs as compared to lowland forests. The position of the belt of maximum precipitation in the mountains depends on the degree of aridity in the adjacent forelands. In case of dry lowlands, the belt of highest rainfall inputs increases, e.g. between 2000 and 2300 m on the southern escarpment of Mt. Kilimanjaro, with rainfall up to 3000 mm a^{-1} (Hemp 2002). In wet foreland areas, maximum precipitation rates occur at much lower elevations, e.g. at 30 m near Debundscha on the western escarpment of Mt. Cameroon with rainfall up to 10,000 mm a^{-1} (Fraser et al. 1999). Further factors affecting the intensity, amount and duration of upslope precipitation are barrier width, slope steepness and updraft speed.

While the principle of the "belt of maximum precipitation input" (Lauer 1976) is valid for most tropical mountains and results from convective airflows, one little known exception must be mentioned: the "Andean Depression of Huancabamba", stretching from Cuenca in southern Ecuador southwards to Chachapoyas, northern Peru, i.e. from around 3 to 7° S. Here, humid air masses of the NE and SE trade winds are pressed by superimposing strong easterlies against the eastern escarpment of the relatively low Cordillera (max. 3900 m). In this case, rainfalls increase continuously up to the crestline, which is almost permanently hidden by cloud caps. This involves leeward foehn effects and resembles the linear precipitation gradient typical for extratropical mountains and resulting from prevailing frontal advection currents (Emck 2007).

Some attributes of tropical climates are important genetic triggers for soil processes. While low mineral fertility due to limited CEC-rates are a well known phenomenon of lowland rainforests, altitudinal gradients of nutrient rates are less investigated. Analyses of specific petrographic substrates indicate increasing contents of three-layer clay minerals and consequently higher CEC-rates due to less intensive chemical weathering in cooler elevations (e.g. Sierra Madre de Chiapas; Richter 1993). However, also in mid-altitudes with maximum precipitation inputs leaching processes might cause low mineral fertility. This is valid for the Cordillera de San Francisco near Loja, southern Ecuador, where a species-rich upper timberline in perhumid environments at 3200–3400 m coincides with extremely acid soils low in cations (Schrumpf et al. 2001). According to the hypothesis that nutrient competition enhances plant diversity in the tropics (Fränzle 1994, Kapos et al. 1990, Woodward 1996), superior competitors might not be able to dominate effectively in this case. In contrast, monotypic timberlines at higher locations such as those formed by *Pinus hartwegii* in México and Guatemala around 4000 m, or by

Polylepis spp. up to 4800 m in the Central Andes, are associated with cold, semi-humid to semiarid climates without mayor effects on soil nutrient leaching.

In many tropical mountain forests the organic soil horizon and (partly dead) root layer gain thickness with altitude. Similar to mighty raw humus layers under coniferous stands of high elevations, this increase indicates a marked below-ground shift in C allocation. Again, adverse soil chemical conditions due to cool temperatures and water logging must be considered decisive factors for the accumulation of necromass as well as the high belowground biomass produced by fine roots (Leuschner et al. 2007, see also chapter 8 of this volume).

Distribution and differentiation of tropical mountain forests

The above-mentioned features of tropical timberlines imply that tropical mountain forests differ considerably within their various distribution areas. On this account, only an approximate definition can be derived from the following climatic and physiognomic features: Tropical mountain forests occur in tropical temperate up to cold altitudinal belts (*tierra templada* to *tierra fria*) and are characterized by a multiple stand structure and an uneven canopy layer. They range from evergreen broadleaved forests with many tree species and a rich vascular as well as cryptogamic epiphytic layer under perhumid conditions to species-poor open woodlands under semiarid conditions. In the first case, trees reach heights up to 40 m while in the latter individuals higher than 15 m are scarce. Independent of the duration of the dry season most tree species possess evergreen scleromorphous leaves. However, two to three months of water deficit may cause some tree species to become leafless for a short period.

Major differences exist with respect to altitudinal range and stand density of tropical mountain forests (Fig. 1). They extend from dense mountain rainforests stretching across more than 2000 m of relative height under (semi-)humid conditions to open woodlands closely confined to an elevated belt of few hundred meters in mountains arising from perarid surroundings. The outer right-hand part of Fig. 1 corresponds to the southwestern escarpment of Mt. Cameroon and the western escarpment of the Andes in Colombia. The example following to the left reflects the outer Andean escarpments of northern Ecuador, the Cordillera de Talamanca in Costa Rica and Panamá, the eastern slopes of the Altos Cuchumatanes in Guatemala and the high mountains in insular Southeast Asia and New Guinea. The middle segment relates to most of the high mounts in eastern Africa and to many of the remaining sections in the Latin American mountain ranges. Finally, the two left-hand types fit the Simien and Balé Mountains of Ethiopia and the western escarpment of the Peruvian Andes. Tropical desert mountains like those in the northern Atacama or the Sudanese Jebel Marra harbour only relic tree stands.

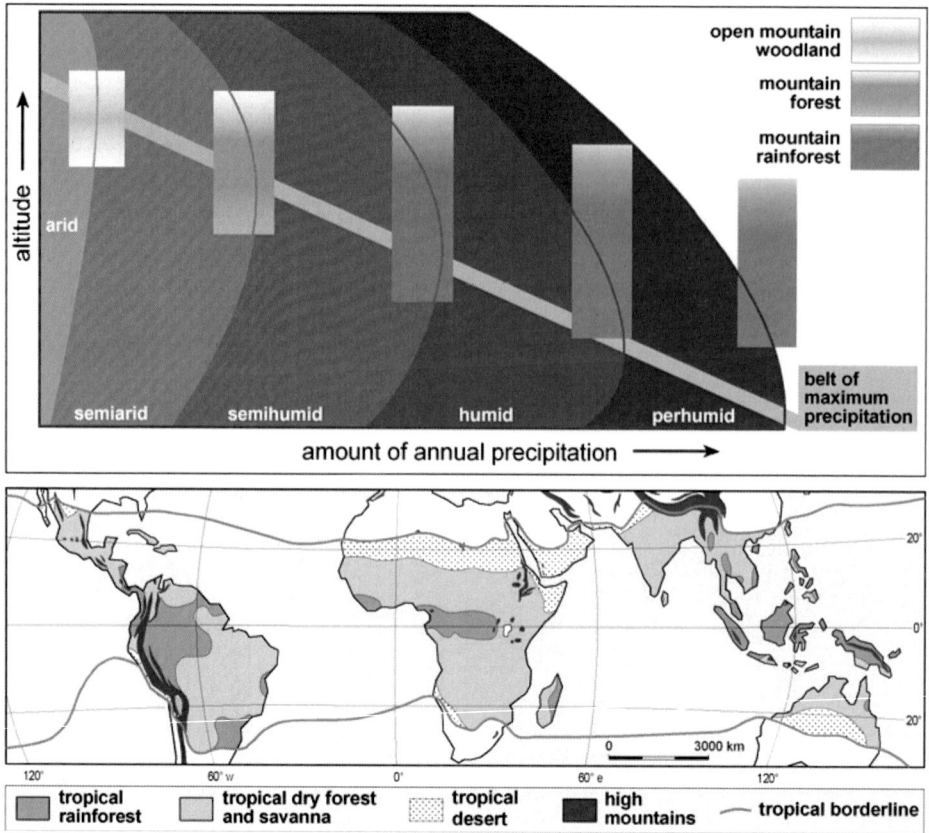

Figure 1. Vertical range of mountain forest types in relation to various tropical precipitation gradients (top) and potential distribution of mountain forests in relation to arid, semiarid/semihumid and humid zones (below).

A notable phytogeographic difference emerges by comparing the upper vegetation belts of the inner and outer tropics (Fig. 2; after Richter 2003). Those of the outer tropics show a dominance of holarctic plant elements, especially in Nepal (Annapurna). Coniferous mountain forests are prominent in the upper montane belts and holarctic elements dominate in the alpine meadows. In contrast, the lower vegetation belts in Nepal are controlled by tropical elements, due to less pronounced thermal seasons. The reason for the commonness of holarctic elements in the high mountain forests of Nepal is the several months lasting snow coverage, which never occurs on the highest peaks of southern Mexico. Situated 14° latitudes further south, the upper mountain rainforests in the latter area are inhabited by a mixture of holarctic tree genera and tropical as well as subantarctic shrubs, herbs and, especially, epiphytes.

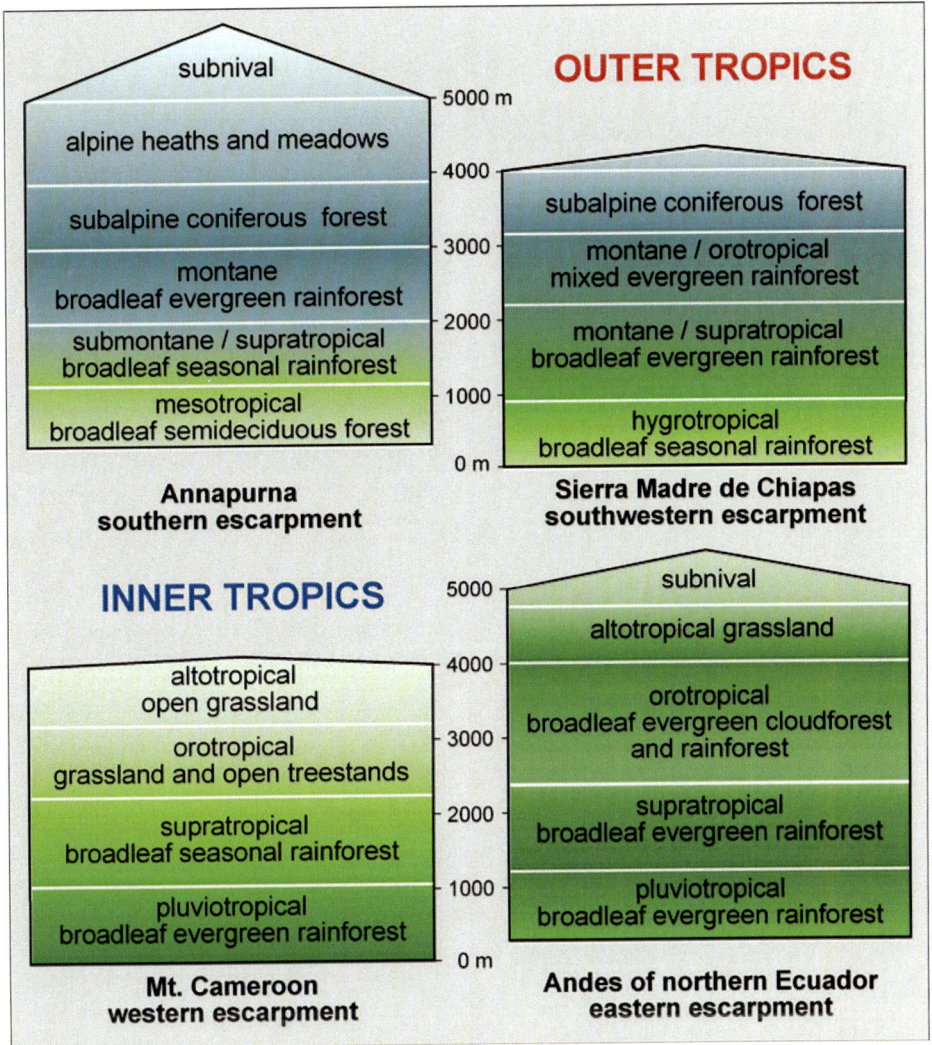

Figure 2. Major altitudinal belts of plant formations in two mountain areas within the outer tropics (top; left Nepal, right southern México) and the inner tropics (below).

The vegetation belts of the inner tropics, both of low and of high elevation, are characterized by tropical and temperate elements (Cleef 1979, Gradstein et al. 1989). Physiognomic differences occur relative to the vertical extension of mountain forests. Differences are small on Mt. Cameroon where the timberline is low (2200–2400 m) due to frequent bush fires, but much more pronounced in the northern Andes, where several tree species of *Polylepis* and *Gynoxys* reach 4000 m due to their special ecophysiological constitution. While the forest at the low

Photoplate 1. (a) montane mixed forest with *Tsuga dumosa* at 2600 m near Dobang, southern escarpment of Annapurna, Nepal. (b) supratropical broadleaved forest with up to 40 m high *Quercus corrugata*, *Sloanea ampla* and *Sterculia mexicana* at 1500 m below Boquerón, western escarpment of the Sierra Madre de Chiapas, México. (c) tree fern-gap with *Cyathea manniana* around 1500 m near Crater Lake, western escarpment of Mt. Cameroon. (d) relic elfin forest stand with *Weinmannia mariquitae* at 3600 m near Saraguro, Nudo de Loja, southern Ecuador.

timberline of Mt. Cameroon is a cloud forest made of up to 30 m tall trees with large stems, fog-induced elfin forests at high timberlines consist of short, often multi-trunked and densely branched, stunted trees.

Finally, it should be noted that mountain rainforests made up of evergreen, medium-sized trees, mossy trunks and a relatively dense understorey are a surprisingly widespread feature of tropical mountains. Even in very dry regions such as the Simien Mountains in northern Ethiopia they occupy a discrete belt on the north-eastern escarpment between about 2200 to 2500 m, even though the rainy season is short and water input limited. The latter observation is suggestive of the high importance of water capture by fog-stripping as a common and decisive driving force in these ecosystems.

Floristics and physiognomy of tropical mountain forests

It is much more difficult (if not impossible) to construct a system of altitudinal belts for tropical mountain forests than for extratropical ones (Fig. 3). For the most part, this is due to greater richness of tree species in the tropics. In the Alps, a rather clear division exists between the lower montane belt consisting of about three dominating tree species (e.g. *Abies alba*, *Fagus sylvatica*, *Acer pseudo-platanus*), and the upper montane and subalpine belts dominated by only two (e.g. *Picea abies*, *Larix decidua*). Between the resulting two vegetation belts a relatively narrow ecotone forms the transition zone. This rather clear pattern contrasts with the chaotic random distribution of the species-rich tree communities in tropical mountain forests, where ecotones and discrete vegetation borders are not easily visible, if at all (Bach 2004).

In spite of this, noticeable floristic, physiognomic and physiological differences exist between the lower and upper parts of inner tropical mountains stretching across a wide vertical range. In humid areas lichens, mosses, hepatics and pteridophytes abound, albeit at different degrees. The ratio of coverage of mosses: lichens, for instance, shows a rather strong increase of mosses with elevation. Among ferns, tree ferns (Cyatheaceae) play an important role in most of the lower mountain forest areas while in the upper reaches filmy ferns (Hymenophyllaceae) profit of high air humidity due to the cool, wet microclimate in the low forest stands, closed by a dense canopy with bowl-formed crown structures. Among flowering plants, in the lower mountain rainforests Lauraceae, Rubiaceae and Piperaceae are frequent associates among trees and shrubs while Araceae and Begoniaceae constitute regularly occurring herbs. Melastomataceae and Peperomiaceae are prominent members of the entire mountain forest system. In upper mountain forests, Asteraceae and Ericaceae gain in importance. Bambusoideae often create impermeable understories in elfin-forest stands with Arundinariinae being predominant in the Paleotropics and Chusqueinae in the Neotropics. Some herbaceous genera are typical members of tropical mountain rainforests and are rare in surrounding ecosystems, e.g. *Impatiens* in Africa and *Bomarea* in the Andes.

Trees and epiphytes are usually the most frequent life forms in tropical mountain rainforests. The most prominent structural elements of plant formation with increasing elevation tree structure change from macro-phanerophytic to nano-phanerophytic, correlating with a decrease of the aboveground biomass. Similarly, the average size of vascular epiphytes decreases with altitude yet not necessarily their species number. On the contrary, since large epiphytes occupy more space than smaller ones, small epiphytes may generate greater species richness. The coverage of foliicolous epiphytes, however, seems to decrease with altitude.

According to Gradstein & Pócs (1989) the upper limit of the occurrence of epiphylls (3000 m) is determined by night frost events. Possibly, changes in the microstructure of the epicuticular waxes of the host leaves, their hydrophobic quality and distinguished UV-reflectance towards higher reaches contribute to the loss of these life forms too (to be investigated). Likewise, leaf sizes decrease with elevation to avoid extreme transpiration rates and radiation maxima. So-called "global super-irradiance" up to 1832 W m^{-2} observed during long-term measurement at 3400 m in southern Ecuador (Emck & Richter, unpubl.) is reflected by red-bluish colors of fresh sprouting leaves. Here, ferns of the genus *Blechnum* and some ericads (e.g., *Cavendishia, Gaultheria, Macleania*) seem to produce enhanced contents of anthocyanin pigments against radiation stress. Decreasing leaf sizes correlate with a decline of the leaf area index (LAI), which might be controlled by reduced soil nitrogen supply (Moser et al. 2007). Finally, a change from dominance of wooden lianas or hemi-epiphytes and trunk climbers in lower mountain forests to herbaceous climbers in higher areas is observed along elevational gradients in tropical mountain forests.

Reasons for high biodiversity in tropical mountain forests

The latest map of the global species diversity of vascular plants (Barthlott et al. 2005) emphasizes tropical mountain areas as the world´s most important diversity hotspots. To illuminate the outstanding role of tropical mountain forests for plant species richness, an approach taking into account different scale sizes (ecozone, landscape, mountain, belt, patch scale; Grüninger 2005) serves as a useful tool.

At the ecozone scale several general (paleo-)ecological factors are valid for most parts of the humid tropics, among them greater effective evolutionary time, i.e. greater evolutionary speed at tropical temperatures and longer contemporary evolutionary time under relatively constant thermal conditions (Rohde 1992). The constant input of "accidentals", i.e. species that are poorly suited for different habitats, is another factor artificially inflating species numbers in the tropics (Stevens 1989). The "seasonal variability hypothesis" considers the limited temperature fluctuations as advantage, by which low-latitude species sustain less violent competitive evolutionary pressures compared to those of high latitudes (Runkle 1989). Better than the frequently cited "intermediate disturbance hypothesis" (Grime 1973) Huston´s "dynamic equilibrium model" (1994) fits tropical ecosystems by assuming an equilibrium of varying growth rates of individual species, low intensities of interspecific competition and low frequencies as well as intensities of disturbances.

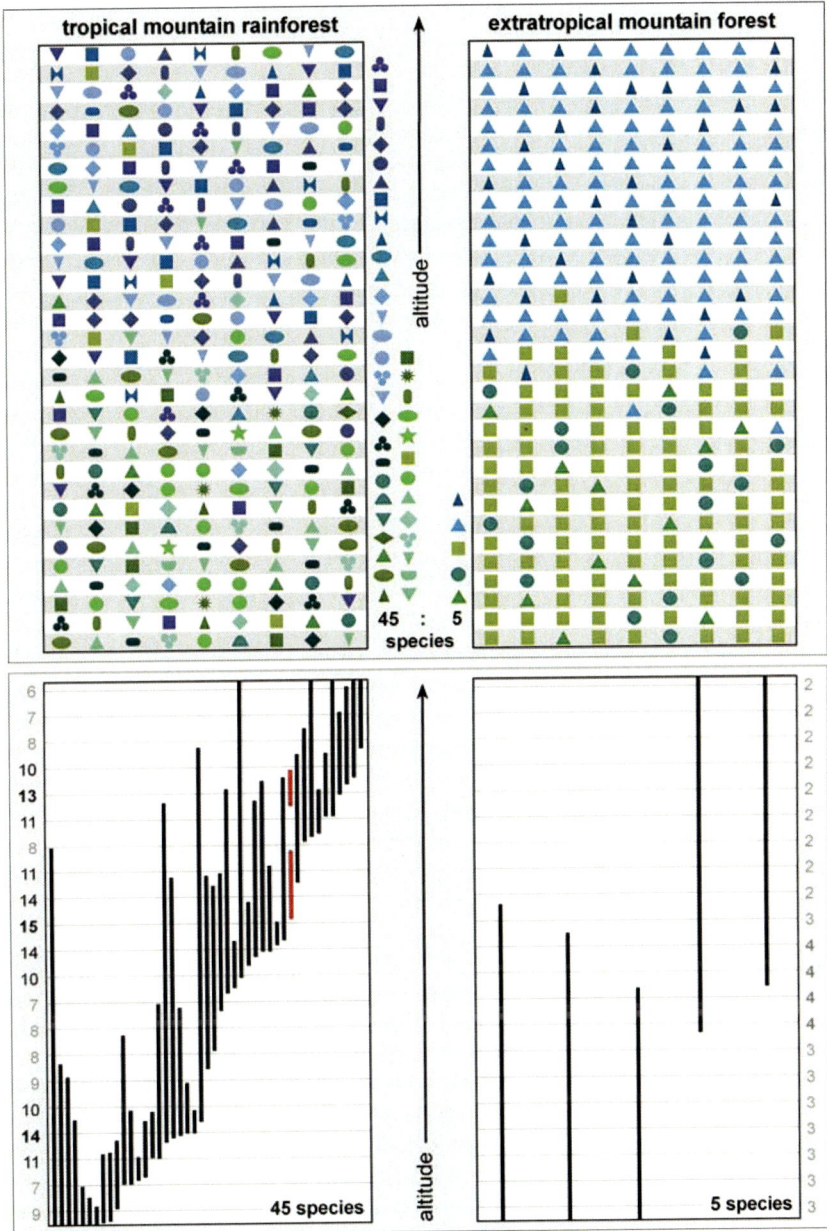

Figure 3. Tropical type of random distribution (left) versus extratropical type of stepwise differentiation (right); the arrangement of tree species within the two contrasting altitudinal mountain forest transects is symbolized by icons (top) and columns (below). Note the unequal diversity patterns of tree species in tropical (high richness) and extratropical (low richness) regions.

Furthermore, infertile soils may increase plant diversity in the tropics (Woodward 1996) because strong competitors are not able to dominate very quickly allowing many species to coexist in a non-equilibrium state. Finally, the "rainfall-diversity hypothesis" (Gentry & Dodson 1987), stating that highest species richness is achieved in wet forests with high and largely non-seasonal annual rainfalls, seems to be a driving force behind the diversity of epiphytes and possibly also terrestrial ferns (Kessler 2001, see also chapter 3 of this volume).

The latter hypothesis explains very well the high diversity of tropical mountain rainforest ecosytems at the landscape scale. According to Barthlott et al. (2005), highest estimated plant species richness is in the Borneo ecoregion including Mt. Kinabalu followed by nine mountain areas in Central and South America. The extraordinary genetic resources of most lowland areas also affect the species number of higher elevations. Furthermore, wide-stretching mountain chains such as the neotropical Cordilleras are a meeting point of tropical lowland and mountain taxa as well as cooler extratropical ones. Additionally, the uplift of mountain chains causes an extensive evolutionary radiation of many paleo-elements and further adds to the high diversity of mountain forests. Lowland immigrant species in mountain habitats may also become founder populations for adaptive radiation. Valley regions and adjacent slopes are most favorable to such a development of endemic species swarms ("evolutionary explosion", Gentry 1982). Gentry (1986, 1995) also emphasizes habitat fragmentation and variation in climatic conditions as decisive triggers for high levels of endemism in tropical uplands. Especially modern families with extensive seed production and short regeneration cycles, such as epiphytic bromeliads and orchids, react genetically flexible due to their generative power to compensate for mortality imposed by habitat patchiness and disturbance (Benzing 1987).

At the mountain scale, the complex topography of many tropical mountains is of fundamental relevance for the petrographic, edaphic, climatic, phytogeographic and hence diversity structures. In many places an extraordinary variety of wet and dry habitats is located in close proximity due to extremely broken terrain with differently exposed slopes, for example in the convergence zone of the three Columbian cordilleras, in the Nepalese Annapurna Massif and on the Hawaiian Island of Kauai. The upper part of the Cordillera Real in southern Ecuador receives more than 6000 mm a^{-1} of rainfall while only 30 km further west rain-shadow conditions with frequent sunshine and dry katabatic winds result in dry conditions with less than 400 mm a^{-1} (Richter 2003). Here, the intricate orography with its complex climate causes close proximity of xeric to hygric vegetation types, contributing to high species richness in mountain forests. Many taxa in the wet forests are physiologically highly plastic while those of the drier mountain woodlands are genetically more stable (Gentry 1982).

At the belt scale, mountain chains generally have a higher biodiversity due to additional taxonomic input from cooler climates in the higher altitudinal belts (e.g. Nagy et al. 2003, Ozenda 2002) (Fig. 4). Evolutionary advantages arise due to the

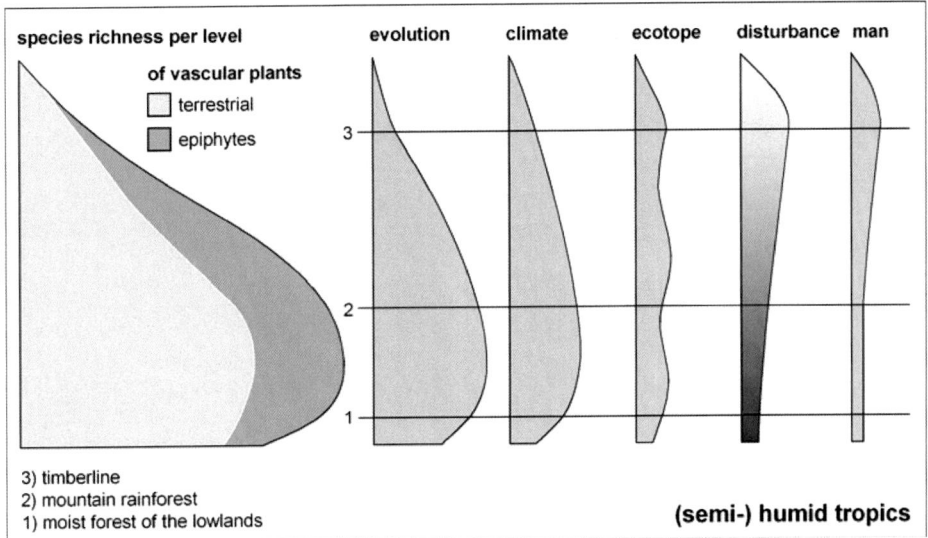

Figure 4. The importance of biotic and abiotic factors driving species richness along a characteristic tropical mountain gradient consisting of a semi-humid foreland, a perhumid intermediate zone and a semi-humid to semi-arid upper zone.

accumulation in the transitional (lower) mountain belt of plant species migrating upslope or downslope during periods of strong climate changes. Evolution is benefited by relatively stable climatic conditions in mild and humid altitudinal zones. While the parameters "evolution" and "climate" are of special importance for the mountain forest belt (Fig. 4), the ecotope, disturbance or human impacts may be of greater relevance for various mountain belts. Furthermore, topography is important at belt scale since an intricate terrain provides multiple combinations of aspects, slope angles, small valleys and ridges. Hence, ecotonal interactions fostering species richness are not only restricted to the transition zone but are a significant feature between the lowest and highest part of the slopes of individual catchments (Beck et al 2007, Oesker et al. 2007). The resulting combination of species distribution renders it difficult or even impossible to recognize specific plant communities. Different habitats are often marked by smooth transition zones (Fig. 2). While this is true for undisturbed areas, disturbances create a patchwork of different succession stages with their pioneer plant communities contributing to local species richness (e.g. Böhmer & Richter 1997, White & Jentsch 2001).

The smallest scale is restricted to micro-habitats and patches of only few square meters. Micro-habitats and their diversity structures play an important role with regard to epiphyte-rich forest and shrub environments due to an often extremely high niche complexity. In northern Ecuador, Freiberg and Freiberg (2000) showed that fewer epiphytic taxa occur below 2000 m due to the larger size and

Photoplate 2. (a) narrow gap caused by old trees toppled over at Meili Xue Shan at around 2500 m in the marginal outer tropics near Dejin, northern Yunnan. (b) gap caused by wildfire on Mt. Chirripó in the Cordillera de Talamanca near timberline at ca. 3400 m, Costa Rica. (c) gap caused by forest elephants with secondary growth of *Afromomum* spec. at around 1600 m near Crater Lake, western escarpment of Mt. Cameroon. (d) gaps by natural landsliding in the RBSF research area at around 2300 m in the Cordillera Real between Loja and Zamora, southern Ecuador.

stronger reproductive competition of some abundant bromeliad and orchid species. Furthermore, mountain forests at lower altitudes have a lesser coverage of moss impeding the establishment of vascular epiphytes on the naked bark. In contrast, the extended moss layers on branches and trunks of trees in the elfin forests at high elevations enhance the settlement of vascular epiphytes. On the other hand, the cooler climate at higher elevations reduces the number of vascular

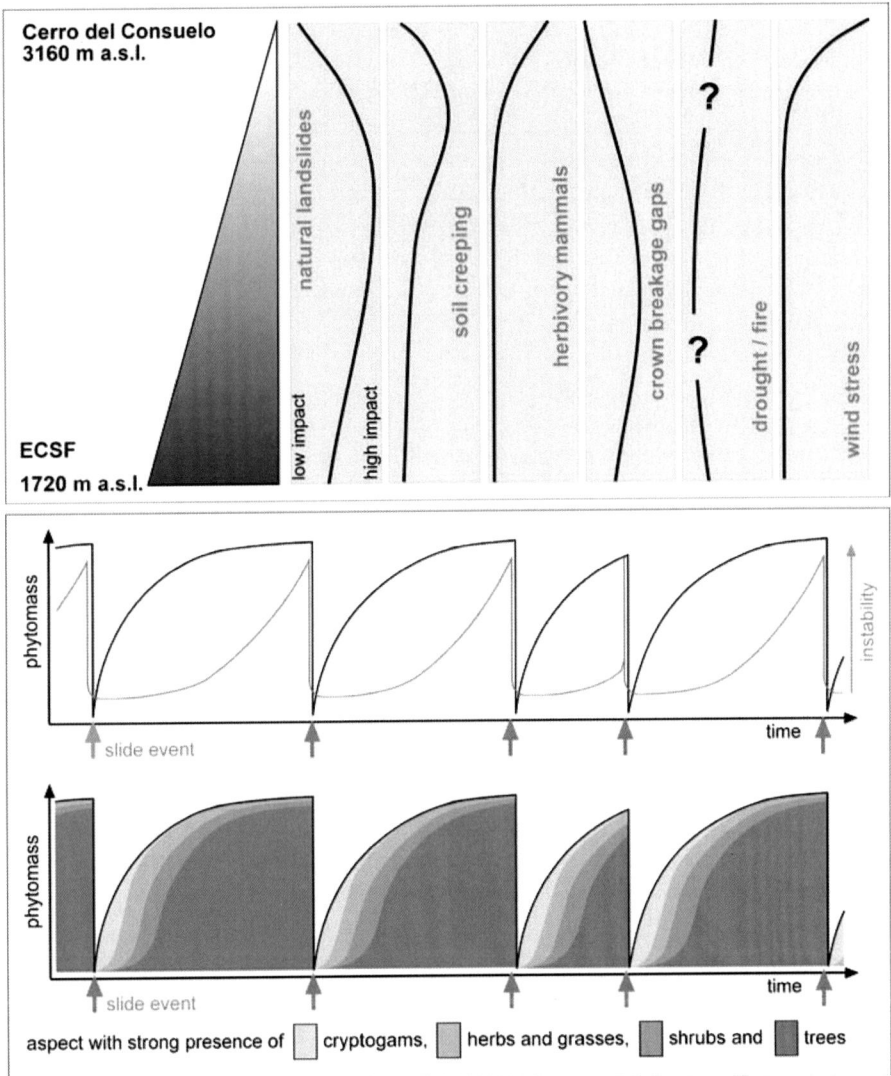

Figure 5. Relative frequency of the most important disturbance types along an elevational gradient in the Reserva Biológica San Francisco (RBSF), Cordillera Real, southern Ecuador (top); sliding processes coupled with phytomass development as a hypothetic framework for forest regeneration in perhumid tropical mountains (below). ECSF: Estación Cientifica San Francisco.

epiphyte species. Additionally, the space for epiphyte communities becomes reduced with decreasing forest height.

Independent of scale, mountain forests are subject to various disturbance factors (crown breakage, landslide, drought, fire, wildlife) (Fig. 5). As a result, various successional plant formations ranging from moss and lichen carpets, herbaceous and bamboo stands, fern thickets and scrub to tall forest may coexist. The resulting mix may be optimal for genetic exchange. At the patch scale, crown breakage or partial die-back of forest trees has little effect on the establishment of heliophile pioneer species since limbs of neighbouring trees may close small forest gap openings very soon. Instead, medium and large-sized disturbances may facilitate the invasion of these pioneers considerably.

Landslides, for example, show various patterns of successional plant communities. Thus, plant assemblages on the upper or lower ends of the slide may differ substantially depending on soil conditions. This habitat heterogeneity coupled with high degree of patchiness in areas of frequent landslides supplementary explains the high species richness of many tropical mountain forest ecosystems. This is especially true for mountain regions characterized by perhumid conditions, metamorphic, limestone or sedimentary rocks, or frequent earthquake activity. In response to these abiotic pressures, the critical threshold for inducing landslides depends on soil mechanical (cohesion and angle of internal friction) and soil hydrological parameters in combination with the destabilization of the water-soaked organic layer by the forest load (van Asch et al. 1989). As illustrated in Fig. 5, the spatio-temporal pattern of landslides governs a mosaic cycle of the forest ecosystem enabling the recovery of adapted species (e.g. Mueller-Dombois 1995, Böhmer & Richter 1997, White & Jentsch 2001).

Tropical mountain forests within the highland-lowland interactive system: importance and threats

The natural sliding processes described in the previous chapter are important driving forces of the evolution and ecology of tropical mountain landscapes. A very different, highly damaging role is played by rotational slides and mud flows resulting from human impact. The negative effects of these human-induced mass movement processes are operative at different levels, ranging from individual species to interactive landscape systems. In this chapter, the negative impact at the interactive landscape level will be described based on our investigations in a tropical mountain region of southern Chiapas (Richter 1993 and 2000).

Within this interactive highland-lowland system, area is affected by various mutual fluxes. For example, the agricultural conversion from orchards to dry farming in the coastal lowlands during the 1970s and 80s has resulted in an upward flux of overheated air masses, less fog in the mountains and hence a major regional climate change. Drastic effects may also result from changes in coffee cultivation, from shaded to non-shaded systems. Measurements show significant increases in stand precipitation, potential evaporation and infiltration rates, and a significant decrease in water holding capacity (Fig. 6).

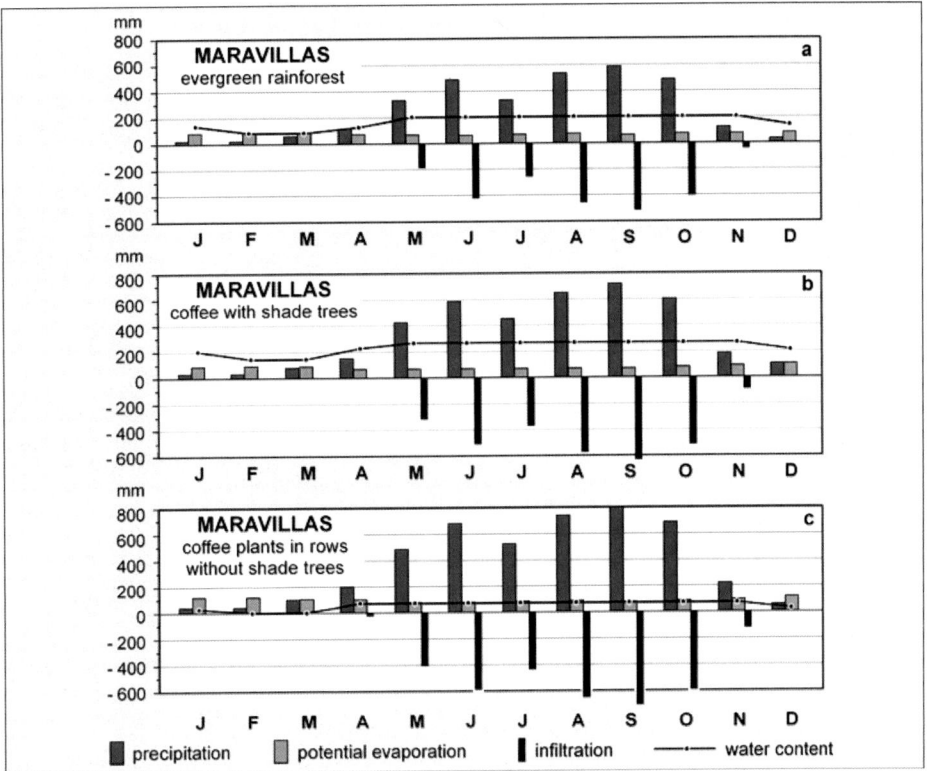

Figure 6. Water regime of the Rio Cuilco catchment around Finca Maravillas, southern Chiapas (ca. 750 m) based on climate data from 1965-1975, modeled infiltration and soil moisture analysis (Schmiedecken 1981). (a) Natural conditions before coffee cultivation. (b) Traditional coffee plantation with shade trees. (c) Coffee plantation without shade trees (coffee plants in rows).

One of the consequences is a hundred to thousand fold raise in soil loss when compared with land under traditional cultivation (Hagedorn 1995) and a dramatic change in stream drainage. The most damaging result is a higher run off rate during the wet period with exuberant river discharges, as exampled by deadly floods during the passage of hurricane Stan in September 2005.

The latter hurricane event, which led to sedimentations of catastrophic magnitude in the foreland of the mountain chain, is one of many examples caused by the destruction of the water and soil reservoirs of the tropical mountain forest ecosystems. Protection of the impacted lowland area by artificial dams and retention ponds, and securing the slide areas by wire nets, does little to solve the problem in the long term. Instead, there is an urgent need of improving the local knowledge and appreciation of the ecological values of forests in these highly vulnerable tropical mountain areas.

References

Bach K (2004) Vegetationskundliche Untersuchungen zur Höhenzonierung tropischer Bergregenwälder in den Anden Boliviens. Dissertation, Universität Göttingen. Görich & Weiershäuser, Marburg

Barthlott W, Mutke J, Rafiqpoor MD, Kier G, Kreft H (2005) Global centres of vascular plant diversity. Nova Acta Leopoldina NF 92, 342: 61-83

Beck E, Makeschin F, Haubrich F, Richter M, Bendix J (2007) The Ecosystem (Reserva Biológica San Francisco). In: Beck E, Bendix J, Kottke I, Makeschin F, Mosandl R (eds) Gradients in a Tropical Mountain Ecosystem of Ecuador. Ecological Studies 198. Springer, Berlin, Heidelberg, New York, pp 1-14

Benzing DH (1987) Vascular epiphytism: taxonomic participation and adaptive diversity. Annals of the Missouri Botanical Garden 74: 183-204

Böhmer HJ, Richter M (1997) Regeneration of plant communities – an attempt to establish a typology and a zonal system. Plant Research and Development 45: 74-88

Cleef AM (1979) The phytogeographical position of the neotropical vascular páramo flora with special reference to the Colombian Cordillera Oriental. In: Larsen K, Holm-Nielsen L (eds) Tropical Botany. Academic Press, London, New York, pp 175-184

Emck P (2007) A Climatology of South Ecuador - With special focus on the major Andean ridge as Atlantic-Pacific climate divide. University of Erlangen (urn:nbn:de:bvb:29-opus-6563), 275 pp. http://www.opus.ub.uni-erlangen.de/opus/volltexte/2007/656/pdf/PaulEmckDissertation.pdf

Emck P, Richter M (unpubl) The upper threshold of global shortwave irradiance in the troposphere derived from field measurements in tropical mountains

Fränzle O (1994) Thermodynamic aspects of species diversity in tropical and ectropical plant communities. Ecological Modelling 75-76: 63-70

Fraser P, Hall JB, Healey JR (1999) Climate of the Mount Cameroon region. University of Wales, Bangor (SENR Publication Number 16), 60 pp

Freiberg M, Freiberg E (2000) Epiphyte diversity and biomass in the canopy of lowland and montane forests in Ecuador. Journal of Tropical Ecology 16: 673-688

Gentry AH (1982) Neotropical floristic diversity: phytogeographical connections between Central and South America, Pleistocene climatic fluctuations, or an accident of the Andean orogeny? Annals of the Missouri Botanical Garden 69: 557-593

Gentry AH (1986) Endemism in tropical versus temperate communities. In: Soulé ME (ed) Conservation biology: the science of scarcity and diversity. Sinauer Associates, Sunderland, Massachusetts, pp 153-181

Gentry AH (1995) Patterns of diversity and floristic composition in Neotropical Montane Forests. In: Churchill SP, Balslev H, Forero E, Luteyn JL (eds) Biodiversity and conservation of neotropical montane forests. New York Botanical Garden, Bronx, pp 103-126

Gentry AH, Dodson CH (1987) Diversity and biogeography of neotropical vascular epiphytes. Annals of the Missouri Botanical Garden 74: 205-233

Gradstein SR, Van Reenen GBA, Griffin D (1989) Species richness and origin of the bryophyte flora of the Colombian Andes. Acta Botanica Neerlandica 38: 439-448

Gradstein SR, Pócs T (1989) Bryophytes. In: Lieth H, Werger MJA (eds) Tropical Rain Forest Ecosystems. Ecosystems of the World, 14B. Elsevier, Amsterdam, pp 311–325

Grime JP (1973) Control of species density in herbaceous vegetation. Journal of Environmental Management 1: 151-167

Grüninger F (2005) Scale dependant aspects of plant diversity in semiarid high mountain regions. An exemplary top-down approach for the Great Basin (USA). Passauer Schriften zur Geographie 21, Passau, 143 pp

Hagedorn A (1995) Untersuchungen zur Bodenerosion in der Kaffeeregion Soconusco, Südmexiko. Mitteilungen der Fränkischen Geographischen Gesellschaft 42: 165-181

Hemp A. (2002) Ecology of the pteridophytes on the southern slopes of Mt. Kilimanjaro I - Altitudinal distribution. Vegetatio 159: 211-239

Huston MA (1994) Biological Diversity. The coexistence of species on changing landscapes. Cambridge University Press, Cambridge, UK

Kapos V, Pallant E, Bien A, Freskos S (1990) Gap frequencies in lowland rainforest sites on contrasting soils in Amazonian Ecuador. Biotropica 22: 218-225

Kessler M (1999) Plant species richness and endemism during natural landslide succession in a perhumid montane forest in the Bolivian Andes. Ecotropica 5: 123-136

Kessler M (2001) Pteridophyte species richness in Andean forests in Bolivia. Biodiversity and Conservation 10: 1473-1495

Lauer, W (1976) Zur hygrischen Höhenstufung tropischer Gebirge. Biogeographica 7: 169-182

Leuschner Ch, Moser G, Bertsch C, Röderstein M, Hertel D (2007) Large altitudinal increase in tree root/shoot ratio in tropical mountain forests of Ecuador. Basic and Applied Ecology 8: 219-230

Moser G, Hertel D, Leuschner Ch (2007) Altitudinal change in LAI and stand leaf biomass in tropical montane forests – a transect study in Ecuador and a pan-tropical meta-analysis. Ecosystems 10: 924-935

Mueller-Dombois D (1995) Biological diversity and disturbance regimes in island ecosystems. Ecological Studies 115: 163-175

Nagy L, Grabherr G, Körner Ch, Thompson BA (2003) Alpine biodiversity in Europe. Ecological Studies 167: 1-477

Oesker M, Dalitz H, Günter S, Homeier J, Matezki S (2008) Spatial heterogeneity patterns – a comparison between gorges and ridges in the upper part of an evergreen lower montane forest. In: Beck E, Bendix J, Kottke I, Makeschin F, Mosandl R (eds)

Gradients in a Tropical Mountain Ecosystem of Ecuador. Ecological Studies 198. Springer, Berlin, Heidelberg, New York, pp 267-275

Ohl C, Bussmann RW (2004) Recolonisation of natural landslides in tropical mountain forests of Southern Ecuador. Feddes Repertorium 115: 248-264

Ozenda P (2002) Perspectives pour une géobiologie des montagnes. Presses Polytechniques et Universitaires Romandes, Lausanne, 208 pp

Richter M (1993) Ecological effects of inappropriate cultivation methods at different altitudes in the Soconusco Region / Southern México. Plant Research and Development 37: 19-44

Richter M (2000) The Ecological Crisis in Chiapas: A Case Study from Central America. Mountain Research and Development 20: 332-339

Richter M (2001) Vegetationszonen der Erde. Klett-Perthes, Gotha, 426 pp

Richter M (2003) Using epiphytes and soil temperatures for eco-climatic interpretation in southern Ecuador. Erdkunde 57/3: 161-181

Rohde K (1992) Latitudinal gradients in species diversity: the search for the primary cause. Oikos 65: 514-527

Runkle JR (1989) Synchrony of regeneration, gaps, and latitudinal differences in tree species diversity. Ecology 70: 546-547

Schmiedecken W (1981) Humidity and cultivated plants – an attempt at parallelizing zones of humidity and optimal locations of selected cultivated plants in the tropics. Plant Research and Development 17: 45-57

Schrumpf M, Guggenberger G, Schubert Ch, Valarezo C, Zech W (2001) Tropical montane rain forest soils - development and nutrient status along an altitudinal gradient in the south Ecuadorian Andes. Die Erde 132: 43-59

Stevens GC (1989) The latitudinal gradient in geographical range: how so many species coexist in the tropics. American Naturalist 133: 240-256

Troll C, Paffen K (1964) Die Jahreszeitenklimate der Erde. Erkunde 18: 1-28

Van Asch ThWJ, Deimel MS, Haak WJC, Simon J (1989) The viscous creep component in shallow clayey soil and the influence of tree load on creep rates. Earth Surface Processes and Landforms 14: 557-564

White PS, Jentsch A (2001) The search for generality in studies of disturbance and ecosystem dynamics. Progress in Botany 62: 399-449

Woodward CL (1996) Soil compaction and topsoil removal effects on soil properties and seedling growth in Amazonian Ecuador. Forest Ecology and Management 82: 197-209

Biodiversity and Ecology Series (2008) 2: 25-33
The Tropical Mountain Forest – Patterns and Processes in a Biodiversity Hotspot
edited by S.R. Gradstein, J. Homeier and D. Gansert
Göttingen Centre for Biodiversity and Ecology

Tropical mountain forest dynamics in Mata Atlantica and northern Andean biodiversity hotspots during the late Quaternary

Hermann Behling

Albrecht von Haller Institute of Plant Sciences, Department of Palynology and Climate Dynamics, University of Göttingen, Untere Karspüle 2, 37073 Göttingen, Germany, Hermann.Behling@bio.uni-goettingen.de

Abstract. Case studies from Morro de Itapeva in southeastern Brazil and El Tiro in the southeastern Ecuadorian Andes document our understanding of tropical mountain vegetation and fire dynamics in biodiversity hotspots during the late Quaternary. Palaeoecological data obtained by pollen and charcoal analysis of radiocarbon-dated sediment cores indicate that present-day mountain forest areas in the study areas were largely covered by grassland vegetation during glacial times. Isolated, small páramos in the Andes and *campos de altitude* areas in southeastern Brazil were connected to larger areas. Mountain rainforests developed during the Late-glacial and in particular during the Holocene. Connection and disconnection of different plant populations may have been an important factor triggering speciation during the Quaternary.

Introduction

The mountain ecosystems of the Neotropics such as mountain rainforests, cloud forests, *Araucaria* forests, páramos and *campos de altitude* (high elevation grassland), especially those of the northern Andes and the Mata Atlantica in eastern Brazil, harbour the most biodiverse ecosystems on planet earth. Yet, little is known about the causes of this high biodiversity. We are especially ignorant of the role history has played in the development of these hotspots of species endemism and richness, despite the importance of understanding the dynamics of the landscape for management and conservation. Palaeoecological studies from the southern Ecuadorian Andes and the Serra do Mar and Serra da Mantiqueira in eastern Brazil provide interesting insights on vegetation, plant diversity, fire and climate dynamics during the late Quaternary (Behling 1997, Niemann & Behling 2008). Important questions addressed include: 1) How was the development and dynamics of tropical mountain ecosystems? 2) Since when do the present-day tropical mountain ecosystems exist? 3) How sensible are tropical mountain ecosystems to climate change and fire? 4) Since when and how strong have tropical mountain ecosystems been influenced by human activity?

Research areas

Southeastern Brazil. We sampled a peat bog (S 22°47', W 45°32') at 1850 m elevation in a small valley basin, about 2 km east of the highest peak of Morro de Itapeva in the Serra da Mantiqueira mountain range of São Paulo State. The Serra do Mar and Serra da Mantiqueira mountain ranges run parallel to the Atlantic coast and form an orographic barrier to the moist easterlies from the Atlantic. The atmospheric circulation of SE Brazil is dominated by the South Atlantic anticyclone, a semi-permanent high pressure system which transports moist tropical air masses to land in an eastern and northeastern direction during the whole year. Orographic precipitation on the Atlantic side of the coastal Serra do Mar amounts to about 4000 mm. The Serra da Mantiqueira, located further inland, receives less rainfall. The climate of the town Campos do Jordão (1600 m) near the study site is mesothermic semihumid, with mean annual rainfall of 1560 mm and a 4 months low rainfall period (Nimer 1989).

Vegetation types in the study area were described by Hueck (1966) and Ururahy et al. (1983). Species-rich *campos de altitude* occur on rounded hill tops at elevations above 1800 m (Fig. 1). The tree line (1800 m) occurs at much lower elevation than in the Andes (Safford 1999a, 1999b, 2001), due to the lower summits and rather dry environments caused by strong winds, thin soil layer and water runoff (Hueck 1966).

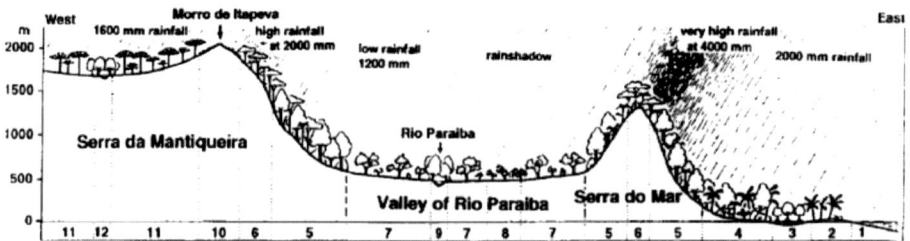

Figure 1. Transect through present-day vegetation from the Atlantic Ocean to Serra da Mantiqueira, southeastern Brazil (adapted after Hueck 1966). **1,** Atlantic ocean. **2,** dune and restinga vegetation. **3,** mangrove. **4,** Atlantic rainforest in the lowland. **5,** Atlantic rainforest on the lower slopes. **6,** cloud forest. **7,** semideciduous forest. **8,** cerrado. **9,** floodplain forest. **10,** high elevation grassland. **11,** *Araucaria* forest. **12,** *Podocarpus* along the rivers.

In addition, anthropogenic impacts such as fires and deforestation have played a role. Depressions and floodplain areas in the highlands between 1400–1900 m are occupied by *Araucaria* forests. Atlantic rainforest occurs on moist slopes between 500 and 1500 m, cloud forest on the Atlantic-facing slopes between 1500–2000 m. "Islands" of cerrado vegetation (savanna, savanna forest, woodlands) occur in the Rio Paraiba valley (Hueck 1956). Since about 150 years the study area is influenced by low-intensity pastural activity.

Southeastern Ecuadorian Andes. Sampling was carried out in a small bog at El Tiro pass (S 03°50'25.9", W 79°08'43.2"), at 2810 m elevation in the eastern Cordillera of the southern Ecuadorian Andes, near to the road from Loja to Zamora. The local climate is influenced by warm moisture-laden air from the Amazon lowland, which collides with cold mountain air masses. This produces much rainfall in the eastern Andean cordilleras, with highest precipitation rates on the western slopes. A short drier period lasts from December until March. Vegetation types in the study area include lower and upper mountain rainforest at 1800–2150 m and 2100–2650(–2750) m, respectively. Subpáramo is present at altitudes between 2800 and 3100 m and páramo from 2900 to 3400 m (Lozano et al. 2003). The timber line in the Loja region is at about 3200 m or about 800 m lower than in central or northern Ecuadorian Andes (Richter & Moreira-Muñoz 2005).

Methods

We studied the vegetation and fire history of the late Quaternary of southeastern Brazil and the southeastern Ecuadorian Andes using pollen and charcoal analysis. Pollen and spores from sediments such as peat deposits can generally be identified to family, often to genus and sometimes to species level. They represent the vegetation of the surroundings of the core location, reflecting vegetation dynamics caused by past climate variability as well as land use changes by forest clearing and use of cultivated plants during the past. Samples from sediment cores were treated by standard pollen analytical methods (Faegri & Iversen 1989) and pollen data were illustrated as pollen percentage diagrams. The rate of charcoal particles in sediment deposits reflects the frequency of past fires.

Results

Southeastern Brazil. A 35,000 yr old core from the Morro da Itapeva mountains (Fig. 2) shows large areas of *campos de altitude* and almost complete absence of *Araucaria* forest, mountain rainforest and cloud forest during 35,000–17,000 yr BP,

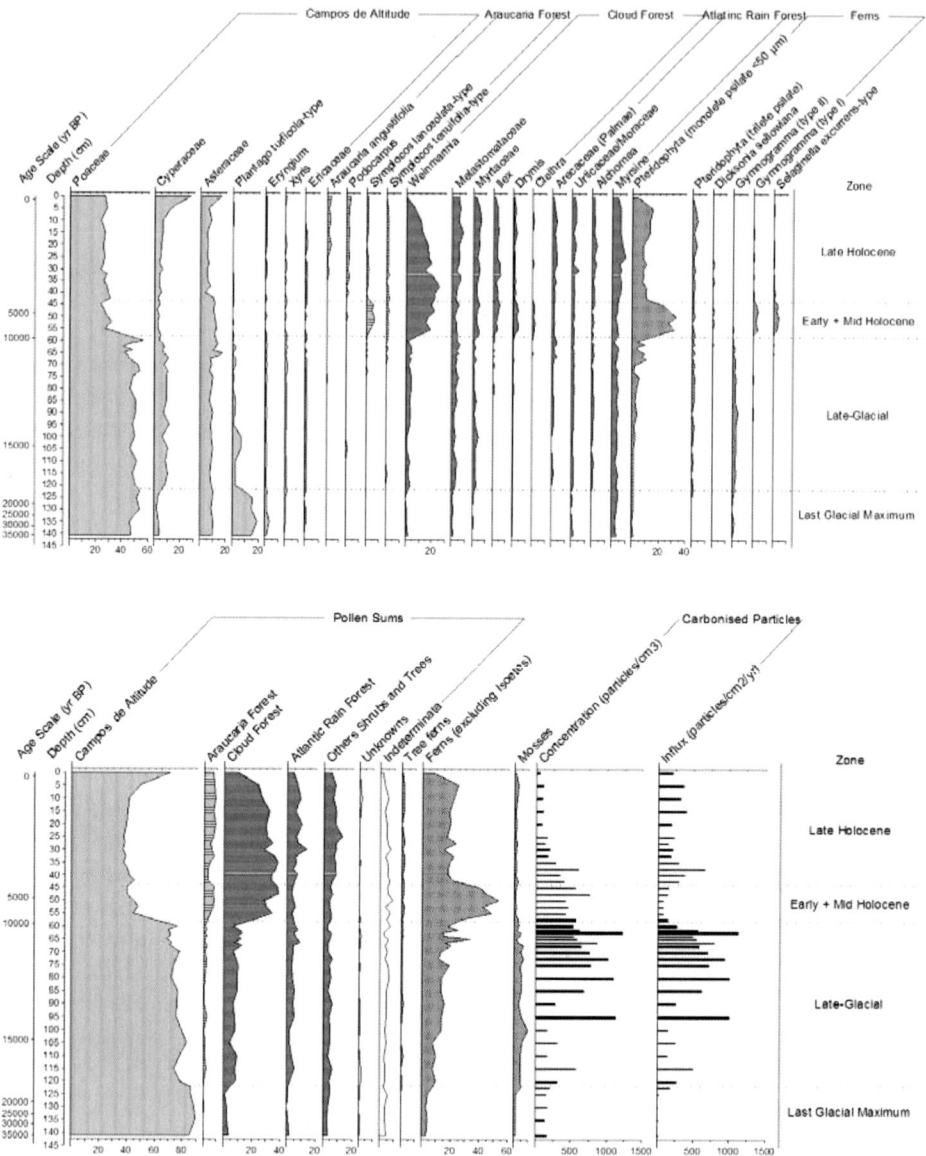

Figure 2. Percentage (*above*) and summary pollen diagram (*below*) of Morro de Itapeva in the Campos do Jordão region of Serra de Mantiqueira in southeastern Brazil. Radiocarbon dates are in uncalibrated years Before Present (yr BP) (from Behling 1997).

indicating a markedly cooler and drier climate than today. Evidence of *Araucaria*, *Podocarpus*, *Drymis* and narrow cloud forest and rainforest belts in the lower mountain region suggest a slightly warmer and moister climate between about 17,000–10,000 yr BP than during the Last Glacial Maximum (LGM) period between ca. 26,000 and 17,000 yr BP.

At the beginning of the Holocene, cloud forest developed close to the Morro da Itapeva site, reflecting a probably warm and moist climate on seaward (east-facing) slopes. Pollen data indicate the full development of Atlantic mountain rainforest. During the early and mid Holocene, the rare occurrence of *Araucaria* and *Podocarpus* and other associated taxa suggest that the climate on the highland plateau was drier than today. In the late Holocene after 3000 yr BP, frequencies of *Araucaria* and *Podocarpus* increased while *Symplocos* decreased, indicating a increase in highland moisture. The presence of *campos de altitude* pollen taxa during the Holocene indicates that the uppermost parts of the Morro de Itapeva mountains remained treeless. Significant parts of the grassland vegetation in the upper regions of the Campos do Jordão region can be considered as natural. However, during the last ca. 150 years grass pollen increased while tree pollen decreased, reflecting human activity in the upper mountain region. Charcoal records from the Morro de Itapeva core show that fires were rare during the LGM and more frequent during the Late-glacial period (i.e. before arrival of man) than during the Holocene (Behling 1997).

Southeastern Ecuadorian Andes. The 17,000 yr BP old core from the El Tiro pass area, which is presently covered by a shrubby subpáramo vegetation, provides important insights on past vegetation, climate and fire history in the region (Fig. 3). The record documents that grass páramo covered the region during glacial times and that the treeline had markedly shifted to lower elevations. The extent of the treeline depression has still to be identified by records from lower elevations. The high frequency of *Plantago rigida* suggests wet climatic conditions during the Late-glacial period. During the early and mid Holocene from about 8000 to 3000 yr BP the grass páramo was almost completely replaced by upper mountain rainforest, reflecting drier and warmer climatic conditions than today. The warmer conditions are reflected in the stronger decomposition of the organic deposits as compared with those of the late Holocene. Since the last 3000 years the area is dominated by shrubby subpáramo.

The charcoal record shows that fires were rare during last glacial and early Holocene period, until 8000 yr BP. Later on, fires became common on the slopes of El Tiro. Three periods with increasing fire intensity have been identified: 8000–3000 yr BP, 2700–1800 yr BP and 1000–600 yr BP. Fire frequency decreased during the last ca. 600 years. The increased fire frequency during the wetter late Holocene suggests that fires were mostly of anthropogenic rather than of natural origin such as lightening. It is assumed that as a consequence of the increased use

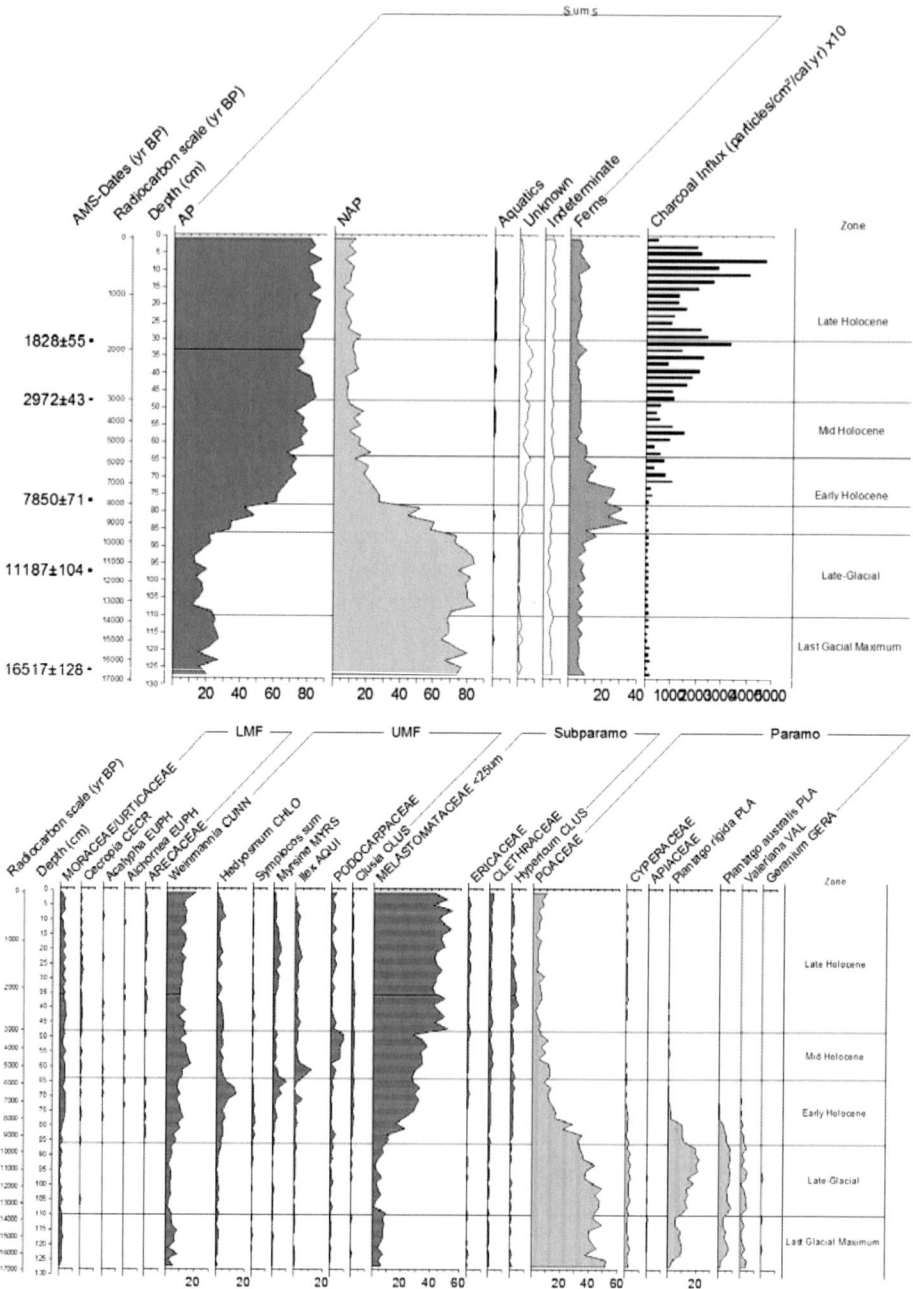

Figure 3. Percentage (*above*) and summary pollen diagram (*below*) of El Tiro in the southeastern Andes of Ecuador (from Niemann & Behling 2008).

of fire, for hunting purposes and by slash and burn cultivation in the drier valleys (e.g. in the Loja area), fires spread into the mountains during drier phases of the year. There is evidence that fire influenced the vegetation composition of El Tiro as indicated by a decrease of common fern taxa including tree ferns. The results clearly indicate that mountain ecosystems are quite sensitive to natural and anthropogenic impacts. Additional investigations should confirm our first results and may allow for environmental reconstructions at the local and regional scale.

Discussion and conclusions

Our studies in the mountains of southeastern Brazil and Ecuador document markedly different vegetations during glacial times as compared with present-day ones. Large areas of mountain forest were replaced by grassland and the treeline had shifted at least 800 to 1000 m downwards. Grasslands was much more extended and currently isolated ones were connected forming larger areas, both in the Andes and in eastern Brazil. The connection and disconnection of isolated grassland areas during Quaternary times must have been an important factor triggering speciation in these tropical mountain regions.

During the LGM the Atlantic mountain rainforest was apparently reduced to a narrow belt at the foot of the Serra do Mar mountain range, in the coastal lowlands, and did not occur further inland to the area between the Serra do Mar and Serra da Mantiqueira where it is found today (Behling & Negrelle 2001, Behling et al. 2002). There, the vegetation consisted of a mosaic of different cerrado types from open grassland (*campo limpo*) to closed dry forest (*cerradão*), probably influenced by fire, and subtropical gallery forests in valleys and on moist floodplain areas. The gallery forest was composed of *Araucaria* forest taxa in the south and tropical semideciduous forest taxa in the north (Behling & Lichte 1997). These forests can be considered as refugia for *Araucaria* forest taxa, which moved during the Late-glacial period to higher elevations, where they occur today. Climatic conditions during the LGM were cold and dry, presumably due to increased frequency of polar cold fronts, and suffered from long dry seasons. The grasslands (*campos*) of the southern Brazilian highlands, which reflect these cold and dry conditions, extended further to the southern part of SE Brazil. Frost sensitive cerrado trees (Silberbauer-Gottsberger et al. 1977), on the other hand, may have existed only in the northern part of southeastern Brazil, where frosts were not as frequent.

In the El Tiro region in southeastern Ecuador, mountain rainforest taxa were very rare or absent during the LGM. Sizable populations occurred only at lower elevations. Climatic conditions were cold and relatively wet, and areas above 3000 m were probably glaciated. During the Late-glacial and, especially, the early Holocene, mountain rainforest developed replacing the former grassland vegetation. The composition of the early to mid Holocene vegetation was different from that of the late Holocene. During the early-mid Holocene upper mountain rainforest was predominant, suggested by frequent occurrence of *Hedyosmun*, Podocarpaceae, *Myrsine* and *Ilex*, and the less frequent occurrence of Melastomataceae. The occurrence of the upper mountain rainforest at the coring site and the strong decomposition of organic material during that time suggest relatively warm and also somewhat drier conditions, as also found in southeastern Brazil. In both regions present-day vegetation became established after ca. 3000 yr BP, when climatic conditions became somewhat cooler and wetter. Major vegetation changes after 3000 yr BP have also been recorded in the Amazonia, where lowland rainforest expanded northwards and southwards (Behling & Hooghiemstra 2000, 2001, Mayle et al. 2000). This supports the hypothesis that changes in the climate regime of the Amazon lowland influenced the mountain rainforest of the eastern Andes and eastern Brazil. While El Tiro mountain became covered by upper mountain forest and subpáramo during the Holocene, the upper part of the Morro de Itapeva remained treeless. The latter species- and endemism-rich area is still naturally covered with *campos de altitude* and should be not become subjected to reforestation. On Morro de Itapeva, fires were rare during the LGM and increaed markedly during the Late-glacial period, apparently due to natural causes. During the Holocene natural and/or anthropogenic fires were less frequent. At El Tiro, fires were rare during the LGM and Late-Glacial period and became frequent only after 8000 yr BP, presumably due to anthropogenic causes. During the last 500 years fire frequency decreased, probably due to a decrease of the local human population.

References

Behling H (1997) Late Quaternary vegetation, climate and fire history from the tropical mountain region of Morro de Itapeva, SE Brazil. Palaeogeography, Palaeoclimatology, Palaeoecology 129: 407-422

Behling H, Arz HW, Pätzold J, Wefer G (2002) Late Quaternary vegetational and climate dynamics in southeastern Brazil, inferences from marine core GeoB 3229-2 and GeoB 3202-1. Palaeogeography, Palaeoclimatology, Palaeoecology 179: 227-243

Behling H, Hooghiemstra H (2000) Holocene Amazon rain forest - savanna dynamics and climatic implications: high resolution pollen record Laguna Loma Linda in eastern Colombia. Journal of Quaternary Sciences 15: 687-695

Behling H, Hooghiemstra H (2001) Neotropical savanna environments in space and time: Late Quaternary interhemispheric comparisons. In: Markgraf V (ed) Interhemispheric Climate Linkages. Academic Press, New York, pp 307-323

Behling H, Lichte M (1997) Evidence of dry and cold climatic conditions at glacial times in tropical Southeastern Brazil. Quaternary Research 48: 348-358

Behling H, Negrelle RRB (2001) Late Quaternary tropical rain forest and climate dynamics from the Atlantic lowland in southern Brazil. Quaternary Research 56: 383-389

Faegri K, Iversen J (1989) Textbook of Pollen Analysis. John Wiley and Sons, New York

Hueck K (1956) Die Ursprünglichkeit der brasilianischen "Campos cerrados" und neue Beobachtungen an ihrer Südgrenze. Erdkunde 11: 193-203

Hueck K (1966) Die Wälder Südamerikas. Fischer, Stuttgart, 422 pp

Lozano P, Delgado T, Aguirre Z (2003) Estado actual de la flora endemica exclusive y su distribucion en el Occidente del Parque Nacional Podocarpus. Funbotanica & Herbario y Jardin Botanico

Mayle F, Burbridge R, Killeen TJ (2000) Millennial-scale dynamics of southern Amazonian rain forests. Science 290: 2291-2294

Niemann H, Behling H (2008) Late Quaternary vegetation, climate and fire dynamics inferred from the El Tiro record in the southeastern Ecuadorian Andes. Journal of Quaternary Sciences 3: 203-212

Nimer E (1989) Climatologia do Brasil. IBGE, Rio de Janeiro, 421 pp

Richter M, Moreira-Muñoz A (2005) Climatic heterogeneity and plant diversity in southern Ecuador experienced by phytoindication. Revista Peruana de Biología 12: 217- 238

Safford HD (1999a) Brazilian páramos I. An introduction to the physical environment and vegetation of the campos de altitude. Journal of Biogeography 26: 693-712

Safford HD (1999b) Brazilian páramos II. Macro- and mesoclimate of the campos de altitude and affinities with high mountain climates of the tropical Andes and Costa Rica. Journal of Biogeography 26: 713-737

Safford HD (2001) Brazilian páramos III. Patterns and rates of postfire regeneration in the campos de altitude. Biotropica 33: 282-302

Silberbauer-Gottsberger I, Morawetz W, Gottsberger G (1977) Frost damage of cerrado plants in Botucatu. Biotropica 9: 253-261

Ururahy JCC, Collares JER, Santos MM, Barreto RAA (1983) Vegetação. In: IBGE, Projeto Radambrasil. Rio de Janeiro/Vitória, Vol. 32: 553-623

Biodiversity and Ecology Series (2008) 2: 35-50
The Tropical Mountain Forest – Patterns and Processes in a Biodiversity Hotspot
edited by S.R. Gradstein, J. Homeier and D. Gansert
Göttingen Centre for Biodiversity and Ecology

Diversity and endemism in tropical montane forests - from patterns to processes

Michael Kessler and Jürgen Kluge

Albrecht von Haller Institute of Plant Sciences, Department of Systematic Botany, University of Göttingen, Untere Karspüle 2, 37073 Göttingen, Germany, mkessle2@uni-goettingen.de

Abstract. It has recently been realized that the dominant pattern of plant and animal species richness along tropical elevational gradients is usually hump-shaped, with maximum richness at intermediate elevations (mostly between 500–2000 m). The causes determining these patterns are still poorly understood but most probably involve area, climatic variables such as temperature and humidity, energy availability and ecosystem productivity, historical and evolutionary processes, and dispersal limitation. Distribution patterns of endemic species along tropical elevational gradients are different from those of species richness and usually peak above 2000 m. The patterns of endemism are presumably determined by area (but inverse to species richness), topography, ecoclimatic stability and taxon-specific ecological traits. While most research is directed at documenting patterns of species richness and endemism, interest in the last two decades has shifted increasingly towards explaining these patterns. The development of testable hypotheses and the use of dated molecular phylogenies have led to important recent progress and are promising exciting new insights in the near future.

Introduction

The documentation and explanation of global and regional gradients of species richness is one of the major challenges of ecological and biogeographical research. The latitudinal gradient of increasing species richness from the boreal regions towards the tropics is one of the best documented patterns in ecology, but the causes determining the observed patterns are still discussed controversially (see review in Hillebrand 2004). In addition to this classical richness gradient, patterns of species richness along elevational gradients have also received considerable attention in the last decade (Rahbek 2005, Lomolino 2001). The elevational richness gradient was long believed to mirror the latitudinal gradient because both span a transition from warm to cold climatic conditions, despite the fact that rainfall and humidity patterns vary greatly between these gradient types. Accordingly, species richness was considered to decline monotonically towards high elevations. Rahbek (1995) showed that this perception was the result of an overemphasis on a few studies showing such monotonic declines, but that across nearly all taxonomic groups the majority of studies have hump-shaped richness patterns with maximum richness at some intermediate point of the gradient (Fig. 1).

While the last few years have seen a flood of new studies confirming highest diversity in montane forests for a wide range of organisms (e.g. Bhattarai et al. 2004, Oommen & Shanker 2005, Kluge et al. 2006, McCain 2007, Fu et al. 2007), the mechanisms or processes determining these richness patterns remain largely unexplored. First, because until recently researchers have mostly been concerned with documenting the patterns, which is a difficult undertaking in inaccessible tropical mountains and for highly diverse taxa. Second, testable hypotheses concerning the potential mechanisms were largely lacking (Evans et al. 2005). And third, methodological limitations precluded the testing of a number of hypotheses, especially those regarding evolutionary processes.

In addition to patterns of species richness, endemism has also figured high on the research agenda of biologists working in tropical montane forests because 1) these forests harbor an astounding number of such locally distributed species, and 2) endemism is of great importance for nature conservation strategies and management.

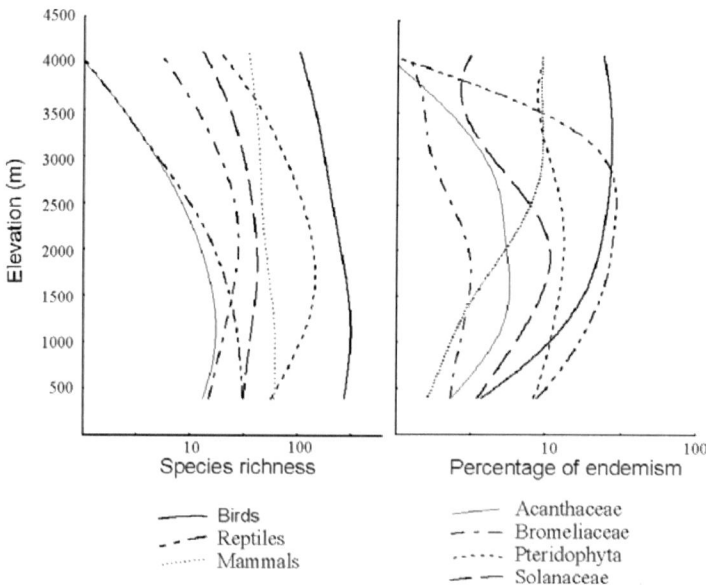

Figure 1. Elevational patterns of species richness (left) and relative endemism (percentage of endemic species; right) for three animal and four plant groups along an elevational transect in central Bolivia. Data generated by D. Embert, S.K. Herzog, E. Jensen, M. Kessler, B. Steudel and T. Tarifa. Note the logarithmic scale of the richness and endemism axes.

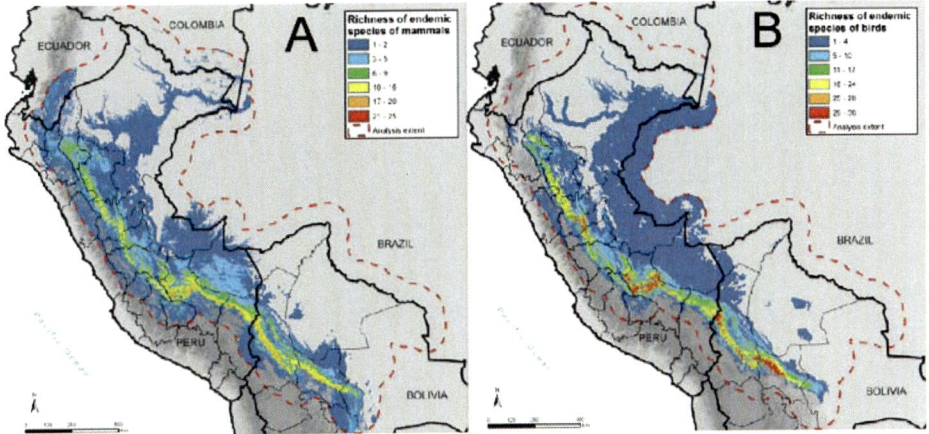

Figure 2. Richness of endemic mammals (A) and birds (B) in the eastern Andes and low-lands of Peru and Bolivia (from Young 2007). Contrary to overall species richness, which peaks in the Amazonian lowlands and the lower foothills, endemic species are concentrated in the Andes. Even there, endemic species are not evenly distributed but are clumped in specific areas, so-called centres of endemism. These are more pronounced for birds than for mammals.

For example, the tropical Andes are considered to be not only the richest and most diverse region on Earth, containing about a tenth of all plant life in less than one percent of the world's land area (www.eoearth.org), but also one of the hotspots of endemism, with at least 15,000 vascular plant species (and possibly many more) occurring nowhere else. However, the geographical distribution of tropical plant endemism is still poorly known, despite several previous efforts to map the ranges of endemic plants and to quantify endemism in tropical plant communities (e.g., Prance 1982, Davis et al. 1997, Pitman et al. 1999). A major problem is the uneven botanical collecting activity over much of the tropics, so that "centres of endemism" frequently correspond to regions of intensive sampling rather than to natural patterns (Nelson et al. 1990). Typically, the proportion of endemic species increases with elevation and peaks above 2000 m, around timberline or even in the alpine zone, and shows a maximum at higher elevations than species richness (Figs. 1, 2). As with patterns of diversity, much research effort is still put into documenting the distribution patterns of endemism whereas the factors determining them remain poorly explored.

The past few years have seen an exciting development in the direction of proposing testable hypotheses determining patterns of species richness and endemism, and research is now moving into a new phase. In the following, we present a brief introduction into some of the most promising directions for future research.

Mechanisms determining patterns of diversity

Area. Going up the mountain, land surface area typically declines. Consequently, patterns of richness in mountains are thought to be strongly influenced by area, in the same manner as islands where large ones have more species than small ones. This "area-effect" might be direct or indirect, overall richness being higher in large areas where more individuals and thus species occur. But even when measuring richness in plots with the same standardized size, the size of the surrounding area in which the plots are nested might have an indirect effect by the population dynamics of bigger or smaller species pools ("echo-effect", Rosenzweig & Ziv 1999). However, the prediction is that lowlands have the highest richness rather than at mid-elevations. Accordingly, area has not been found in any single study to be the main predictor of species richness.

Energy. Although the availability of energy represents the first-ever proposed explanation of diversity gradients (von Humboldt 1808), it is only in the last two decades that contemporary climate-related factors have gained strong attention as putative causes for variations of richness for many organisms (Mittelbach et al. 2001, Rahbek & Graves 2001, Hawkins et al. 2003, Currie et al. 2004). For a number of organisms, but especially for plants, it has been found that species richness at high elevations declines due to reduced temperatures, whereas at low elevations water availability appears to be the limiting factor (Kreft & Jetz 2007).

However, how these factors determine species richness remains unclear. As water availability and temperature are essential for plant growth and drive the energy input in ecosystems, a limitation of one or both of these might influence richness patterns. The 'species-energy theory' states that the more energy is available in an ecosystem, the more species will be able to coexist (Wright 1983). But how to measure energy input? Actual evapotranspiration (AET) tries to combine temperature and rainfall as a surrogate for energy availability and this measure is often positively related to species richness. The same is true for net primary productivity (NPP). The energy assimilated by plants and made available to consumers is most often considered as the driving variable, but in field studies energy measures are mostly restricted to surrogates of NPP (Waide et al. 1999). In most cases indirect variables (e.g., rainfall) or direct variables (e.g., biomass, foliage area) are measured; measures of NPP and richness at the same scale are rare. There is an urgent need for measures of productivity of selected taxa (Mittelbach et al. 2001).

In addition to measurements of energy input and productivity in relation to species richness, the question by which mechanisms they are coupled should be addressed. The latter question is not at all understood, even though several hypotheses exist (Evans et al. 2005). Some of these relate to the dynamics of populations under changing energy conditions. First, it is assumed that more individuals are able to coexist when energy increases. The reason for this assumption is that more populations are lifted above their threshold of viability, thereby reducing the risk of their extinction (*more-individuals-hypothesis*, Srivastava & Lawton 1998). Second, specialized species which use relatively scarce resources may be hampered in low-energy areas because the abundance of such resources may be too low to allow the survival of viable populations (*niche-position-hypothesis*, Abrams 1995). Third, when resource abundance increases, species may focus on their preferred resource types and turn away from less preferred ones, leading to reduced niche breadth in high energy areas. This may lead to reduced niche overlaps and, hence, to lower competition, promoting co-existence and a positive species-energy relationship (*niche-breadth-hypothesis*). As none of these hypotheses have as yet been conclusively tested (Evans et al. 2005), they present exciting opportunities for future research.

History. The structure of ecosystems is predominantly explained by variation in the physical environment. In addition, their today's structure is strongly influenced by historical processes.. Regarding richness patterns, there is recent evidence that local diversity is strongly connected to regional diversity. This raises the question as to the importance of history for local richness patterns (e.g., Ricklefs 2005, Roy & Goldberg 2007). Ecological gradients may show phylogenetic patterns reflecting historical differences between regions as well as inherited traits of clades contributing to the local biodiversity (Ricklefs 2005, Hawkins et al. 2006). For example, the tropics are suggested to harbor a higher diversity across many taxa than temperate regions. Reasons for this are: 1) tropical regions had a greater extent in the historical past (Behrensmeyer et al. 1992), therefore more extant clades are originally tropical; 2) a higher amount of energy in the tropics leads to shorter generation times and thus higher diversification rates (*speciation-extinction-hypothesis*, Cardillo 1999); 3) adaptations necessary to disperse and persist in cold and climatically seasonal regions have evolved only in some taxa (*niche-conservatism-hypothesis*, sensu Wiens & Donoghue 2004). Unfortunately, ecological and historical (phylogenetic) biogeography has experienced a major chasm in explaining diversity patterns in the past 30 years (Harrison & Cornell 2007). Nevertheless, there is no doubt that phylogenetic history is an important component in the analysis of extant biodiversity (Harrison & Grace 2007).

Null models. Another recent development in the study of the origin, maintenance and distribution of biotic diversity has been the consideration of null models and stochastic factors (Gotelli & Graves 1996). The basic idea is that when we take an existing range sizes of species and place them randomly within geographically or ecologically constrained domains, these ranges will overlap in the center of the domain more frequently than at the domain edges, in this way creating a unimodal richness pattern. This is called the *mid-domain effect* (MDE) and has been invoked as an explanation for richness patterns along gradients (see review in Colwell et al.

2004). MDE models have been shown to correlate closely to diversity patterns especially along some elevational gradients (Kessler 2000, Colwell et al. 2004, McCain 2004), but the meaning and implications of such geometric constraint models are hotly debated (e.g., Grytnes 2003, Colwell et al. 2004). The main criticism is that range size frequency distribution, which generates hump-shaped patterns due to the MDE null model, not necessarily exists in the absence of environmental gradients (Hawkins et al. 2005).

On a different spatial scale, Hubbell (2001) has suggested that the occurrence and abundance of species at given sites within a larger ecoregion or habitat is strongly influenced by dispersal limitation. The *Neutral Theory* considers variations in the dispersal ability of species, their population size and immigration (resp. speciation) rate as the crucial factors determining the composition and abundance distribution of species. The relative influence of these factors over deterministic ecological factors varies with scale and study group. At the local scale, dispersal limitation seems the most important factor (e.g., Svenning et al. 2004, Vormisto et al. 2004).

Figure 3. Some of the world's most diverse ecosystems are found on the lower slopes of the Andes, partly due to the large variety of ecosystems. This view of the Cordillera Mosetenes in central Bolivia shows the range from tall forests on alluvial soils in the valley bottom (ca. 750 m) to stunted forests on ridges at ca. 1800 m. But also at a given elevation, as in the foreground, there is a striking difference between the tall forest on the right and the shrubby vegetation on the left. This is caused by different soil conditions. Finally, there are abundant natural disturbances due to landslides.

At the regional scale and across strong ecological gradients, environmental conditions such as soils and climate (*niche assembly model*) are more important (e.g., Condit et al. 2002, Phillips et al. 2003, Tuomisto et al. 2003, Jones et al. 2006) (Fig. 3). Because of the overriding influence of environmental changes along elevational and precipitation gradients, the neutral theory of biodiversity has not been applied to these gradients as contrary to the mid-domain one.

Mechanisms determining patterns of endemism

Endemic species have by definition spatially restricted ranges and are often very scarce within their distributional range, making them a prime target of conservation activities. A main feature of endemic species is that they are distributed non-randomly. They are especially well-represented on oceanic islands (Carlquist 1974) and on "habitat islands" such as mountain ranges (Gentry 1986) (Figs. 1, 2). However, endemic species also show clumped patterns of occurrence in apparently homogeneous habitats, such as Amazonian rain forests (Prance 1982) or humid Andean forests (Fjeldså 1995) (Fig. 2). The latter situations have been interpreted in the light of various historical processes such as glacial cycles, climatic instability and floodplain dynamics (Salo et al. 1986). As a result, the factors invoked as potentially determining patterns of endemism differ noticeably from those determining patterns of species richness.

Area. As with overall species richness, area influences patterns of endemism, but the relationship is inverse. Whereas species richness increases in larger areas, the fraction of endemic species increases in smaller areas. This is because small habitat islands harbor small populations of species that can rapidly diverge genetically and ecologically. This area effect is closely linked to topographic structure (Fig. 4). For example, several recent studies have shown that with increasing area of the elevational belts, percentage endemism decreases (Kessler 2002). Mountains, which are typically relatively small compared to the surrounding lowlands and are often separated from other mountain ranges, support many plant and animal populations that are relatively small and isolated from closely related populations. These isolated populations can diverge genetically and thereby form new, endemic taxa. This has been found in several studies of endemism and is the most commonly cited explanation for elevational patterns of endemism (e.g., Graves 1988, Kruckeberg & Rabinovitz 1985, Ibisch et al. 1996, Sklenář & Jørgensen 1999, Kessler 2000, 2002, Kessler et al. 2001, Kluge & Kessler 2006). The frequently observed decrease of endemism at highest elevations (usually above timberline) can be explained by various processes. In some cases recent mountain uplift is assumed to have provided too little time for speciation, while in other cases Pleistocene glaciations might have lead to the extinction of alpine endemics (Major 1988). In the Andes, the higher geographical connectivity of mountain plateaus allows high-montane species to be widely distributed, whereas species inhabiting the steep, topographically complex slopes have narrow, fragmented ranges and hence tend towards local endemism (Graves 1988, Sklenář & Jørgensen 1999, Kessler 2001a, 2002).

Figure 4. Treeline habitats in the Bolivian Andes are not overly rich in species but support a high number of endemic species (Fig. 2). This view at 3450 m shows stunted forest vegetation as well as the extremely steep and rugged topography that causes species ranges to be narrow and fragmented, favouring the evolution of new species with restricted ranges. The less steep, more accessible slopes on the far left are largely devoid of forest due to regular burning by local inhabitants, which constitutes a high level of threat to many of the endemic species.

Stability. In addition to the general elevational patterns detailed above, there are also regional shifts in the level of endemism. Some areas in the Andes harbor high concentrations of endemic taxa whereas others at the same elevation are inhabited mainly by widespread species. Fjeldså & Lovett (1997) and Fjeldså (2007) linked concentrations of endemic bird and plant species in Africa and South America to low levels of environmental disturbance. The latter explanation was based on the observation that distributions of recently evolved species and old relict ones overlapped. Further support of the latter hypothesis was obtained by remotely sensed data, reflecting ecoclimatic stability in the eastern slopes of the Andes of Ecudaor and Peru (Fjeldså et al. 1999). Ecoclimatic stability not only correlated with patterns of bird endemism but also with the location of past and present centers of human cultures, suggesting that the latter also depended on stable climatic conditions (see also Fjeldså 2007). Ecoclimatic stability may thus be a major factor determining the spatial distribution of plant, animal and human populations in the Andes. While intriguing, the studies by Fjeldså and colleagues were limited to few groups of organisms in few habitat types and did not consider other environmental factors, which might have conceivably influenced bird endemism and human population distribution. Furthermore, it is not known whether the observed correlation does indeed reflect a causal relationship. For example, ecoclimatic stability is linked to special topographic conditions such as funnel-shaped valleys, which are characterized by high environmental heterogeneity (Fjeldså et al. 1999). High levels of endemism in these sites may also have been due to the high habitat variability, enabling range-restricted species with specific habitat requirements to migrate to suitable habitats in close proximity during past climatic changes. In less heterogeneous habitats, these species may have become extinct due to migration constraints.

Taxon specific traits. In addition to the above-mentioned factors, patterns of endemism may differ based on the taxon-specific ecological traits. Such traits involve aspects of reproduction, dispersal, life form, demography, spatial population structure and competitive ability of organisms (Kruckeberg & Rabinovitz 1985, Major 1988, Kunin & Gaston 1993, Kessler 2001 a). For example, in many groups of plants epiphytic species tend to be more widespread than their terrestrial counterparts (Ibisch et al. 1996, Kessler 2000, 2001a, 2002). Since the relative abundance of epiphytic plants varies with elevation, this may well influence the elevational pattern of endemism of a given plant family and override the effect of topography. Such taxon-specific traits have influenced the specific evolutionary history of the organisms, leading to group-specific elevational patterns of endemism. Overall, the most likely explanation of the variety of observed patterns is that there are various separate mechanisms determining endemism, whose signals are only visible at different levels of observation. When we study patterns of endemism at the level of genera or families, taxon-specific traits and evolutionary histories are likely to be the most important explanatory factors. However, when we focus on the whole flora, taxon-specific patterns are blurred and topographical effects become more important.

Disturbance. Patterns of endemism are also influenced by natural and human disturbances. The relationship of plant ranges size to habitat disturbance has been studied only superficially and still lacks a coherent theoretical background (Huston

1994). Along successional vegetation gradients on landslides, plant groups often show an increase in species richness from early to late successional stages and mature forest, and a subsequent decline in senescent forest (e.g., Kessler 1999). In contrast, endemic species are best represented in very early successional stages and in mature forest, perhaps as a result of less stable conditions at mid-succession and due to competitive displacement of endemic species in senescent forest. Along gradients of human disturbance, endemism shows a roughly similar pattern, but with one important exception: in 21 of 25 studied cases, endemic species were somewhat more common in anthropogenically disturbed than in natural forests (Kessler 2001b).

Endemic taxa are frequently given high priority in conservation strategies (Bibby et al. 1992, Williams et al. 1996, Davis et al. 1997) because their small ranges render them particularly vulnerable to habitat loss (Balmford & Long 1994), and because they are assumed to be more susceptible to anthropogenic habitat disturbance than widespread taxa (Andersen et al. 1997). However, the high values of endemism at low levels of human disturbance and at early successional stages observed in our studies in Bolivia suggest that endemic humid montane forest species may not be quite so sensitive to habitat disturbance as currently assumed. This is not surprising considering that montane forests are subject to intensive natural disturbances by landslides and tree falls (Garwood et al. 1979, Kessler 1999) and that many species may therefore be adapted to disturbances.

Conclusions

For many major plant and animal taxa, the global to regional distributional patterns of species richness and to a lesser degree of endemism are now reasonably well known. Much remains to be learned about the patterns of little-studied and inconspicuous taxa (many insect groups, soil organisms, etc.). In addition, local distributional patterns are not well understood for many taxa. While additional surveys are needed to address these deficits, the most important future discoveries are likely to come out of studies on the mechanisms causing these patterns. Important progress has been made in the last years in two fields. First, the development of testable hypotheses allows the transition from descriptive research to hypothesis-driven work. Second, the widespread application of molecular methods will enable to unravel the evolutionary processes leading to the present-day patterns. The combination of these two approaches will be particularly powerful.

A better knowledge of the patterns and causes of species richness and endemism in tropical mountain forests may have important implications for conservation and biodiversity management, especially in view of global change. Thus, increasing endemism with elevation calls for a large number of conservation areas at high elevations, particularly around timberline elevation (Myers et al. 2000) (Fig. 4).

However, the optimal location and design of such a network of timberline reserves is not yet known, because of the paucity of biological data from many timberline regions, the limited overlap of patterns of richness and endemism between different organism groups, and the difficulty of modeling such patterns from environmental data. Clearly, extensive, well-planned additional field work is needed here. Given the rather low species richness above 3000 m, timberline conservation units need not be particularly large in order to representatively cover local vegetation types and to maintain viable species populations. At lower elevations, however, high species richness, low population density of many species (Pitman et al. 1999), comparatively low endemism, and high spatial ecological variability (Fig. 3) call for larger conservation areas.

Important future questions for optimizing habitat and species conservation and management include the following: What is the relationship of biotic community composition, diversity and endemism in upper montane forests to human activities? Which human uses are sustainable at what intensity? Is the conservation of upper montane forests dependent on strict reserves or is it more efficiently conducted in conservation areas subject to human extraction practices?

Acknowledgements. We thank Rodrigo Soria and Bruce Young for figures.

References

Abrams PA (1995) Monotonic or unimodal diversity-productivity gradients, what does competition theory predict? Ecology 76: 2019-2027

Andersen M, Thornhill A, Koopowitz H (1997) Tropical forest disruption and stochastic biodiversity losses. In: Laurance WF, Bierregaard RO (eds) Tropical forest remnants: ecology, management, and conservation of fragmented communities. University of Chicago Press, Chicago, pp 281-292

Balmford A, Long A (1994) Avian endemism and forest loss. Nature 372: 623-624

Behrensmeyer AK, Damuth JD, DiMichele WA, Potts R (1992) Terrestrial ecosystems through time: evolutionary paleoecology of terrestrial plants and animals. University of Chicago Press

Bhattarai KR, Vetaas OR, Grytnes JA (2004) Fern species richness along a central Himalayan elevational gradient. Journal of Biogeography 31: 389-400

Bibby CJ, Collar NJ, Crosby MJ, Heath MF, Imboden C, Johnson TH, Long AH, Stattersfield AJ, Thiergood SJ (1992) Putting biodiversity on the map, priority areas for global conservation. ICBP, Cambridge

Cardillo M (1999) Latitude and rates of diversification in birds and butterflies. Proceedings of the Royal Society of London, B 266: 1221-1225

Carlquist S (1974) Island Biology. Columbia University Press, New York

Colwell RK, Rahbek C, Gotelli NJ (2004) The mid-domain effect and species richness patterns: what have we learned so far? American Naturalist 163: E000-E023

Condit R, Pitman NCA, Leigh EG, Chave J, Terborgh J, Foster RB, Núñez V P, Aguilar S,Valencia R, Villa G, Muller-Landau HC, Losos E, Hubbell SP (2002) Beta-diversity in tropical forest trees. Science 295: 666-669

Currie DJ, Mittelbach GG, Cornell HV, Field R, Guégan J-F, Hawkins BA, Kaufman DM, Kerr JT, Oberdorff T, O'Brien E, Turner JRG (2004) Predictions and tests of climate-based hypotheses of broad-scale variation in taxonomic richness. Ecology Letters 7: 1121-1134

Davis SD, Heywood VH, Herrera-McBryde O, Villa-Lobos J, Hamilton AC (1997) Centres of Plant Diversity, a Guide and Strategy for their Conservation. Vol. 3: The Americas. IUCN-Publ. Unit Cambridge, U.K.

Evans KL, Warren PH, Gaston KJ (2005) Species-energy relationships at the macroecological scale: a review of the mechanisms. Biological Reviews 80: 1-25

Fjeldså J (1995) Geographical patterns of neoendemic and relict species of Andean forest birds: the significance of ecological stability areas. In: S. P. Churchill SP, Balslev H, Forero E, Luteyn JL (eds) Biodiversity and conservation of Neotropical montane forests. New York Botanical Gardens, Bronx, pp 89-102

Fjeldså J (2007) The relationship between biodiversity and population centers: the high Andes region as an example. Biodiversity and Conservation 16: 2739-2751

Fjeldsa J, Lovett JC (1997) Biodiversity and environmental stability. Biodiversity and Conservation 6: 315-322

Fjeldsa J, Lambin E, Mertens B (1999) Correlation between endemism and local ecoclimatic stability documented by comparing Andean bird distributions and remotely sensed land surface data Ecography 22: 63-78

Fu C, Wang J, Pu Z, Zhang S, Chen H, Zhao B, Chen J, Wu J (2007) Elevational gradients of diversity for lizards and snakes in the Hengduan Mountains, China. Biodiversity and Conservation 16: 707-726

Garwood NC, Janos DP, Brokaw N (1979) Earthquake-caused landslides: A major disturbance to tropical forests. Science 205: 997-999

Gentry AH (1986) Endemism in tropical versus temperate plant communities. In: Soulé MA (ed) Conservation Biology: The Science of Scarcity and Diversity. Sinauer Associates Inc. Sunderland, Massachusetts, pp 153-181

Gotelli NJ, Graves GR (1996) Null models in ecology. Smithsonian Institution, Washington, DC

Graves GR (1988) Linearity of geographical range and its possible effect on the population structure of Andean birds. Auk 105: 47-52

Grytnes JA (2003) Species richness patterns of vascular plants along seven altitudinal transects in Norway. Ecography 26: 291-300

Harrison S, Cornell HV (2007) Introduction: merging evolutionary and ecological approaches to understanding geographic gradients in species richness. American Naturalist 170: S1-S4

Harrison S, Grace JB (2007) Biogeographic affinity helps explain productivity-richness relationships at regional and local scales. American Naturalist 170: S5-S15

Hawkins BA, Field R, Cornell HV, Currie DJ, Guègan J-F, Kaufman DM, Kerr JT, Mittelbach GG, Oberdorff T, O'Brien EM, Porter EE, Turner, JRG (2003) Energy, water, and broad-scale geographic patterns of species richness. Ecology 84: 3105-3117

Hawkins BA, Diniz-Filho JAF, Weis AE (2005) The Mid-Domain Effect and Diversity Gradients: Is There Anything to Learn? American Naturalist 166: E140-E143

Hawkins, B.A., Diniz-Filho, J.A.F., Jaramillo, C.A. & Soeller, S.A. (2006) Post-Eocene climate change, niche conservatism, and the latitudinal diversity gradient of New World birds. Journal of Biogeography 33: 770-780

Hillebrand H (2004) On the Generality of the Latitudinal Diversity Gradient. American Naturalist 163: 192-211

Hubbell SP (2001) The unified neutral theory of biodiversity and biogeography. Princeton University Press, Princeton

Huston MA (1994) Biological diversity. Cambridge University Press, Cambridge

Ibisch PL, Boegner A, Nieder J, Barthlott W (1996) How diverse are neotropical epiphytes? An analysis based on the 'Catalogue of the flowering plants and gymnosperms of Peru'. Ecotropica 1: 13-28

Jones MM, Tuomisto H, Clark DB, Olivas P (2006) Effects of mesoscale environmental heterogeneity and dispersal limitation on floristic variation in rain forest ferns. Journal of Ecology 94: 181-195

Kessler M (1999) Plant species richness and endemism during natural landslide succession in a perhumid montane forest in the Bolivian Andes. Ecotropica 5: 123-136

Kessler M (2000) Elevational gradients in species richness and endemism of selected plant groups in the central Bolivian Andes. Plant Ecology 149: 181-193

Kessler M (2001a) Patterns of diversity and range size of selected plant groups along an elevational transect in the Bolivian Andes. Biodiversity and Conservation 10: 1897-1920

Kessler M (2001b) Maximum plant community endemism at intermediate intensities of anthropogenic disturbance in Bolivian montane forests. Conservation Biology 15: 634-641

Kessler M (2002) The elevational gradient of Andean plant endemism: varying influences of taxon-specific traits and topography at different taxonomic levels. Journal of Biogeography 29: 1159-1166

Kessler M, Herzog SK, Fjeldså J, Bach K (2001) Species richness and endemism of plant and bird communities along two gradients of elevation, humidity and land use in the Bolivian Andes. Diversity and Distributions 7: 61-77

Kluge J, Kessler M, Dunn R (2006) What drives elevational patterns of diversity? A test of geometric constraints, climate, and species pool effects for pteridophytes on an elevational gradient in Costa Rica. Global Ecology and Biogeography 15: 358-371

Kluge J, Kessler M (2006) Fern endemism and its correlates: contribution from an elevational transect in Costa Rica. Diversity and Distributions 12: 535-545

Kreft H, Jetz W (2007) Global patterns and determinants of vascular plant diversity. Proceedings of the National Academy of Sciences 104: 5925-5930

Kruckeberg AR, Rabinowitz D (1985) Biological aspects of endemism in higher plants. Annual Review of Ecology and Systematics 16: 447-479

Kunin WE, Gaston JJ (1993) The biology of rarity: patterns, causes, and consequences. Trends in Ecology and Evolution 8: 298-301

Lomolino MV (2001) Elevation gradients of species-density: historical and prospective views. Global Ecology and Biogeography 10: 3-13

Major J (1988) Endemism: a botanical perspective. In: Myers AA, Giller PS (eds) Analytical biogeography. Chapman and Hall, New York, pp. 117-146

McCain CM (2004) The mid-domain effect applied to elevational gradients: species richness of small mammals in Costa Rica. Journal of Biogeography 31: 19-31

McCain CM (2007) Could temperature and water availability drive elevational species richness patterns? A global case study for bats. Global Ecology and Biogeography 16: 1-13

Mittelbach GG, Scheiner SM, Steiner CF (2001) What is the observed relationship between species richness and productivity? Ecology 84: 3390-3395

Myers N, Mittermeier RA, Mittermeier CG, Da Fonseca GAB., Kent J (2000) Biodiversity hotspots for conservation priorities. Nature 403: 853-858

Nelson BW, Ferreira CAC, da Silva MF, Kawasaki ML (1990) Endemism centers, refugia and botanical collection density in Brazilian Amazonia. Nature 345: 714-716

Oommen MA, Shanker K (2005) Elevational species richness patterns emerge from multiple local mechanisms in Himalayan woody plants. Ecology 86: 3039-3047

Phillips OL, Núñez Vargas P, Lorenzo Monteagudo A, Peña Cruz A, Chuspe Zans, M-E, Galiano Sánchez W, Yli-Halla M, Rose S (2003) Habitat association among Amazonian tree species: a landscape-scale approach. Journal of Ecology 91: 757-775

Pitman NCA, Terborgh J, Silman MR, Nuñez V P (1999) Tree species richness in an upper Amazonian forest. Ecology 80: 2651-2661

Prance GT (ed.) (1982) Biological Diversification in the Tropics. Columbia University Press, New York

Rahbek C, Graves GR (2001) Multiscale assessment of patterns of avian species richness. Proceedings of the National Academy of Science 98: 4534-4539

Rahbek C (1995) The elevational gradient of species richness: a uniform pattern? Ecography 18: 200-205

Rahbek C (2005) The role of spatial scale and the perception of large-scale species-richness patterns. Ecology Letters 8: 224-239

Ricklefs, R.E. (2005) Phylogenetic perspectives on patterns of regional and local richness. In: Bermingham E, Dick CW, Moritz C (eds) Tropical rainforest: past, present, and future. University of Chicago Press, pp 16-40

Rosenzweig ML, Ziv Y (1999) The echo pattern of species diversity: pattern and process. Ecography 22: 614-628

Roy K, Goldberg EE (2007) Origination, Extinction, and Dispersal: Integrative Models for Understanding Present-Day Diversity Gradients. American Naturalist 170: S71-S85

Salo J, Kalliola R, Häkkinen I, Mäkinen Y, Niemelä P, Puhakka M, Coley PD (1986) River dynamcis and the diversity of Amazon lowland forest. Nature 322: 254-258

Sklenář P, Jørgensen PM (1999) Distribution patterns of páramo plants in Ecuador. Journal of Biogeography 26: 681-691

Srivastava DS, Lawton, JH (1998) Why more productive sites have more species, an experimental test of theory using tree-hole communities. American Naturalist 152: 510-529

Svenning J-C, Kinner DA, Stallard RF, Engelbrecht BMJ, Wright SJ (2004) Ecological determinism in plant community structure across a tropical forest landscape. Ecology 85: 2526-2538

Tuomisto H, Ruokolainen K, Aguilar M, Sarmiento A (2003) Floristic patterns along a 43-km long transect in an Amazonian rain forest. Journal of Ecology 91: 743-756

Von Humboldt A (1808) Ansichten der Natur mit wissenschaftlichen Erläuterungen. JG Gotta Tübingen, Germany

Vormisto J, Svenning J-C, Hall P, Balslev H (2004) Diversity and dominance in palm (Arecaceae) communities in terra firme forests in the western Amazon basin. Journal of Ecology 92: 577-588

Waide RB, Willig MR, Steiner CF, Mittelbach G, Gough L, Dodson SI, Juday GP, Parmenter R (1999) The relationship between productivity and species richness. Annual Review of Ecology and Systematics 30: 257-300

Wiens JJ, Donoghue MJ (2004) Historical biogeography, ecology and species richness. Trends in Ecology and Evolution 19: 639-644

Williams P, Gibbons D, Margules C, Rebelo A, Humphries C, Pressey R (1996) A comparison of richness hotspots, rarity hotspots and complementary areas for conserving diversity using British birds. Conservation Biology 10: 155-174

Wright DH (1983) Species-energy theory, an extension of species-area theory. Oikos 41: 496-506

Young BH (2007) Distribución de las especies endémicas en la vertiente oriental de los andes en Perú y Bolivia. NatureServe. Arlington, Virginia, USA

Biodiversity and Ecology Series (2008) 2: 51-65
The Tropical Mountain Forest – Patterns and Processes in a Biodiversity Hotspot
edited by S.R. Gradstein, J. Homeier and D. Gansert
Göttingen Centre for Biodiversity and Ecology

Epiphytes of tropical montane forests - impact of deforestation and climate change

Stephan Robbert Gradstein

Albrecht von Haller Institute of Plant Sciences, Department of Systematic Botany, University of
Göttingen, Untere Karspüle 2, 37073 Göttingen, Germany, sgradst@uni-goettingen.de

Abstract. The abundance of epiphytes is one of the striking characteristics of moist tropical montane forests. Our recent studies in South America show that deforestation may lead
to major losses of epiphytes, especially among orchids, bromeliads and ferns. Drought-
intolerant species are more strongly impacted than drought-tolerant ones. Species turnover
along disturbance gradients is high and recovery of the epiphyte communities in regenerating forest is very slow. The epiphyte flora of fifty year old secondary forest is still very
different from that of the natural forest. Alteration of the microclimate is of paramount
importance in affecting epiphyte diversity along disturbance gradients. Because of their
tight coupling to atmospherical conditions, epiphytes are predictably sensitive to global
warming. The recent northward expansion of bryophytes and lichens in Europe is first
evidence of the impact of climate change on epiphytes in temperate regions. In the tropics,
where long-term distributional records are lacking, the possible impact of global warming
on epiphytes has been studied by means of transplantation experiments. The results are
suggestive of significant changes in the composition of the rich epiphytic communities of
the tropical montane forest over time due to climate change.

Introduction

One of the characteristic features of moist tropical montane forests is the abundance of epiphytes or arboreal plants[1], including orchids, bromeliads, aroids, ericads, peperomiads, cacti, ferns, bryophytes, lichens, etc. (Fig. 1). These plants represent some of the most striking life forms of the forest and play an important
role in the hydrology and mineral cycles of the forest ecosystem. By intercepting
moisture and nutrients from the atmosphere, epiphytes may store substantial fractions of the water and nutrient pools of the forest. Up to 44 tonnes of epiphytic

---

[1] Epiphytes = plants growing on trees or other larger plants and using their hosts only for mechanical support. Unlike mistletoes, epiphytes take water and minerals from the air or suspended soil
and do not penetrate the vessels of the host plant; they are not parasites.

biomass (including suspended soil) have been measured in one hectare of montane cloud forest in the Andes of Colombia, constituing about one third of the total biomass of the forest and almost half of the above-ground stocks of nitrogen and phosphorous (Hofstede et al. 1993). The thick epiphyte mats also attract a great variety of animal groups, fungi and micro-organisms seeking moisture, shelter, food and breeding opportunities, and serve as nesting material for birds (Nadkarni & Matelson 1989, Nadkarni & Longino 1990, Yanoviak et al. 2007).

Species richness of epiphytes in moist tropical montane forests is also very high. In the Reserva Biológica San Francisco, a montane forest reserve of approximately 1000 hectares in southern Ecuador where almost 2400 species of flowering plants, ferns, bryophytes and lichens have been recorded (Liede-Schuhmann & Breckle 2008), about one on every two species is an epiphyte. The number of epiphyte species in montane forest is normally much higher than in tropical lowland forest, except in the cloud forests of wet coastal areas where epiphyte diversity may be as high as in the mountains due to the frequency of fog (Gradstein 2006, Gradstein in press).

Why are montane rain forests so rich in epiphytes? Probably the principal reason is the almost permanently moist environment and the lack of night frosts,

Figure 1. The abundance of epiphytes is one of the striking characteristics of tropical montane forests. More than one hundred epiphyte species may occur on a single tree. Reserva Biológica San Francisco, Ecuador, 2000 m. Photograph Florian Werner.

allowing the plants to thrive year round high up on the trees. Since they can normally take up water only when it is raining, the constancy of precipitation is more important for epiphytes than the total amount of rainfall. This is why epiphytes are particularly abundant in cloud forests, where the air is constantly saturated and the leaves are dripping from cloud water. In general, however, plant growth in the exposed, aerial environments is not without hazards of desiccation and in able to withstand large intervals between rain showers, many epiphytes have developed special structures for water retention such as succulent leaves, leaf bulbs, water sacs, leaf tanks, etc. The velamen of epiphytic orchids and the water-absorbing scales of many bromeliads are adaptive structures to enhance rapid water uptake during rain showers. CAM metabolism, or the nocturnal absorption of carbon dioixide while keeping the stomata closed during the daytime, is another strategy of many epiphytes to avoid water loss. A further adaptation is the reduction of leaves and development of photosynthetic stems as seen in the whisk fern, *Psilotum nudum*. Some epiphytic ferns such as *Asplenium nidus*, the nest fern, produce their own soil by collecting leaf litter and detritus in funnels of ascending leaves. Researchers from Cambridge University have recently shown that these leaf baskets are hot spots of diversity and that one nest fern may harbour as many species of animals as the entire host tree (Ellwood & Foster 2004).

It is generally considered that mycorrhizal fungi play a crucial role in the life of plants in nutrient-poor environments and for that reason one would expect abundant presence of mycorrhizal associations in epiphytes. Surprisingly, however, mycorrhizal fungi are less common among epiphytes than among the soil-based flora, with the exception of the orchids and the ericads. The reasons for the paucity of these fungi in epiphytes are unknown although it has been suggested that this may be due to the high fertility of the suspended soil and the failure of propagules of the mycorrhizal fungi to arrive at the positions where epiphytes are growing. The ability of some epiphytes to develop fungal associations when cultivated in the greenhouse indicates that these plants can form mycorrhiza but as epiphytes apparently do not (Bermudes & Benzing 1989, Nadarajah & Nawami 1993).

Epiphytes include many rare and endemic taxa, especially among the orchids. It is estimated that about one third of the flowering plants of Colombia, totalling almost 25,000 species, are endemic to the country and about half of these are epiphytic orchids (Bernal et al. 2007). In fact, there are more species in orchids than in any other group of plants and many are still awaiting discovery (Fig. 2). High endemism is also found in the aroids and the bromeliads; the latter are abundant in the New World tropics but are absent in Africa and Asia. Ferns, bryophytes and lichens, on the other hand, are generally more widespread than the flowering plants and hold fewer endemics due to their dispersal by light, wind-dispersed spores.

The rich diversity of tropical epiphytes is now becoming increasingly endangered by the destruction of the montane forests and their conversion into pastures and other forms of land use. It is estimated that about 90% of the montane forests of the Andes have already disappeared in spite of numerous efforts to conserve these forests and slow down the pace of deforestation. The rapid conversion of these forests poses the question as to the fate of the rich epiphytic communities.

Figure 2. New orchid species of the genus *Teagueia* from Ecuador. Photographs Lou Jost. See also: www.loujost.com

Can they survive in logged, degraded forests or on remnant forest trees in pastures? Are they able to re-establish in plantations or in regrowing secondary forest, and if yes how fast is their recovery? Can management practices be designed to enhance the survival of the epiphyte diversity after forest disturbance, whether selective logging, burning or clear-cutting?

Another question to be addressed is the possible impact of global warming on epiphytes. Because of the tight coupling of the epiphytes to atmospherical conditions, especially moisture supply, one may expect that changes in the local climate have an appreciable impact on the diversity of these plants (Lugo & Scatena 1992, Benzing 1998). How do epiphytes react to local changes in the climate? Can they survive when the amount of moisture in the environment becomes reduced due to the rising temperatures? These are important questions that need to be answered in order to be able to conserve these plants for the future.

Impact of deforestation on epiphytes

The impact of deforestation on epiphyte diversity has been dealt with in several recent studies and the results have been rather controversial. While several authors reported a marked decrease of species richness of vascular epiphytes following forest disturbance (as one would expect), other authors found little or no change. For example, species richness of bromeliads and orchids was significantly reduced in disturbed montane forest areas in Venezuela and northern Ecuador (Dunn 2000, Barthlott et al. 2001), in other areas it was not (Wiliams-Linera 1990, Hietz-Seifert et al. 1996). Reasons for these different results have been unclear and have prompted us to undertake an investigation of the subject. In this chapter I will briefly describe some of the results of our epiphyte work in Bolivia and Ecuador.

Bolivia. We studied epiphyte diversity in natural forest and adjacent young, 10-15 y old secondary forest at two different elevations (600 m, 1600 m) on the eastern slope of the Andes, near La Paz (Acebey et al. 2003, Krömer & Gradstein 2004). By comparing species diversity and microclimate in these habitats, we tried to obtain insight in the responses of epiphytes to forest alteration, their ability of recovery after deforestation and their value as indicators of disturbance. Trees were ascended by means of the single rope climbing technique (Fig. 3) and epiphyte species were collected from the bases of the trunks to the outer portions of the crowns, using a standard protocol for representative sampling (Gradstein et al. 2003). In this way, ca. 250 species of epiphytes (flowering plants, ferns, bryophytes) were found in just one hectare of forest. This is one of the highest species numbers ever recorded and indicates that the montane forests of Bolivia rank among the richest worldwide in terms of epiphyte diversity.

Figure 3. Ascent into the tree for study of epiphytes by means of the single-rope climbing technique. Reserva Biológica San Francisco, Ecuador, 1900 m. Photograph Florian Werner.

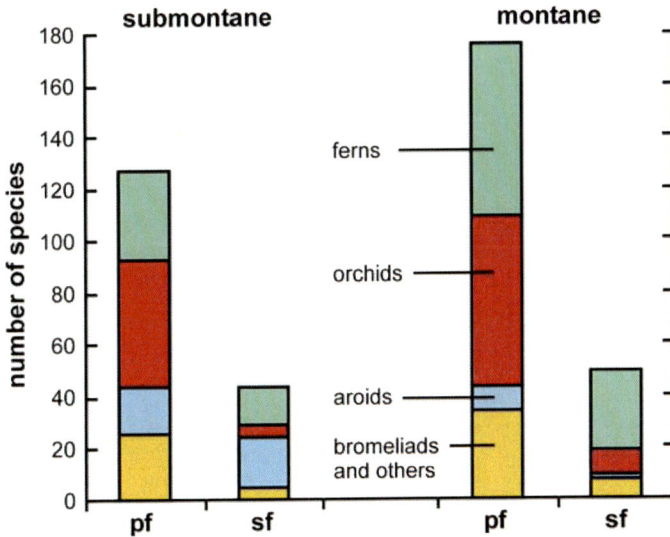

Figure 4. Species richness of vascular epiphytes in primary forest (pf) and young secondary forest (sf) of the submontane belt (600 m) and the montane belt (1600 m) in the Bolivian Andes. Orchid and bromeliad diversity is dramatically reduced in secondary forest at both elevations, that of aroids only in the montane belt (after Krömer & Gradstein 2004)

Deforestation in the study area resulted in major loss of epiphytes: young secondary forests had on average 60-70% fewer epiphyte species than the neighbouring natural forests (Fig. 4). Losses were particularly severe in orchids (90%), bromeliads (70-80%) and filmy ferns (100%). In other plant groups such as fern families Aspleniaceae and Polypodiaceae, and in the liverworts, species richness was not reduced. However, the composition of the epiphytic flora in young secondary forests differed considerably from that in natural forest, indicating that species turnover was high and that many forest species had become lost, especially those of the shaded understory of the forest, the so-called "shade epiphytes". On the other hand, we noted that many epiphyte species of the high canopy of the natural forest, the "sun epiphytes", were able to re-establish on lower parts on the trees in the secondary forest, presumably due to the warmer and drier conditions found there.

The impoverishment of the epiphyte flora in young secondary forest was explained by several factors, three of which may be mentioned here. First, the secondary forest trees had rather smooth bark and were more uniform in architecture than those in natural forest, having little-branched crowns with oblique instead of horizontal branches. This type of crown architecture is unfavourable for the establishment of epiphytes (Ibisch 1996). A second important factor was the drier microclimate in the secondary forest, caused by the more open canopy of the secondary forest and the stronger radiation. Not all epiphyte groups were negatively affected by the decreased air humidity, however. Hemi-epiphytic aroids (starting their life as vines and subsequently becoming epiphytes due to loss of the connection to the soil) were thriving profusely in the well-lit understory of the submontane secondary forests (but not in the montane forest), probably because of their succulent stems and initial connection to the soil. Desiccation-tolerant ferns such as *Pleopeltis astrolepis* and *P. polypodioides*, being well-adapted to rather dry environments, were also common and clearly outnumbered those in the natural forest. A third reason was the lack of a dense bryophyte cover on the secondary forest trees. In mature montane forests, trunks and branches are usually covered by thick moss layers, which provide a suitable substrate for the germination of epiphytic ferns and flowering plants (Nadkarni 1984). The reduced bryophyte cover in the young secondary forest apparently had a negative impact on the diversity of vascular epiphytes in these forests.

Our observations on the dramatic decline of epiphyte richness in secondary forest agreed with earlier findings from Singapore (Turner et al. 1994) but deviated from results from Venezuela (Barthlott et al. 2001) and elsewhere, where species losses were less severe. The lesser decline observed in the Venezuelan study was probably due to the older age (ca. 30 y) of the secondary forest and the inclusion of isolated trees in the study. These trees may be very rich in epiphytes and contribute considerably to the diversity of open, deforested habitats (Hietz-Seifert et al. 1996). Therefore, we were keen to include old secondary forest and isolated trees in our subsequent work.

Ecuador. The Reserva Biológica San Francisco in southern Ecuador, between Loja and Zamora, has been our main fieldwork site for the last seven years, since the establishment of Research Unit 402 "Functionality in a tropical mountain forest" of the German Research Foundation in this area in 2001. Here, we studied epiphyte diversity along a gradient from natural forest across 50 years old secondary forest to isolated trees in pastures and recorded more than 450 species, including about 250 vascular epiphytes (flowering plants, ferns) and over one hundred lichens (Werner et al 2005, Nöske et al. 2008).

As in Bolivia, we observed major floristic changes from natural forest towards open vegetation for all groups of epiphytes. Moreover, we observed the characteristic switch of canopy species towards lower parts on the tree in open habitats. The changes in species richness were very different among the various groups of epiphytes: species numbers declined dramatically (over 80%) in vascular epiphytes, less so (ca. 30%) in bryophytes, and increased by 10-15% towards the open vegetation in lichens (Fig. 5). Especially the tiny microlichens were more numerous in the drier, open habitats than in the forest where they were largely restricted to branches and twigs in the outer portions of the crowns. Interestingly, a similar turnover pattern as in lichens has been observed among moths (Geometridae, Arctiidae), which also show a clear increase in species towards open, disturbed habitats in montane forest areas (Hilt & Fiedler 2005, Hilt et al. 2006).

Figure 5. Abundant growth of epiphytic lichens on an isolated tree in pasture land. Lichens are generally well adapted to withstand long periods of desiccation and are therefore common in open habitats. Reserva Biológica San Francisco, Ecuador, 1900 m. Photograph Florian Werner.

The preference of epiphytic lichens for relatively dry habitats has also been observed elsewhere. In lowland rain forests of Guyana, for example, Cornelissen & Gradstein (1990) found that dry evergreen forests were much richer in lichens than moist evergreen ones. Bryophytes, however, showed just the opposite pattern and had a clear preference for the moist forest. The explanation for the ecological behaviour of the lichens is apparently their ability to withstand long periods of desiccation, longer than in other groups of epiphytes (Frahm 2003).

Within the natural forest, species richness of lichens was significantly lower on shaded lower parts of trunks than in more open canopy habitats. Highest species richness was found on horizontal branches in mid-canopy where growth conditions were enhanced by humus accumulation. Using cluster analysis we were able to distinguish two main groups of epiphyte communities in the forest, those of shaded and of sunny habitats (Nöske 2005). The shade communities occurred on tree bases and on the underside of branches in the forest canopy, the sun communities on the upper sides of canopy branches and on isolated trees. Among the bryophytes and lichens, about ten percent of the species were shade epiphytes, the rest were sun epiphytes or ecological generalists growing in shaded as well as in sunny environments. As in Bolivia, the shade epiphytes were most vulnerable to deforestation.

The results of this study showed that not only forests but also the modified, open habitats may play an important role for the conservation of biodiversity in tropical mountains. As in Bolivia, we found that host tree characteristics, the openness of the canopy and the microclimate in the forest were important parameters determining the responses of the organisms to habitat alteration (Nöske et al. 2008). The very different species composition in the natural and 50 years old recovering forest indicated that the regeneration of the rich epiphyte communities in these mountains forests following disturbance is very slow. These results agree with earlier observations on slow regeneration of epiphyte diversity in montane forests of Costa Rica (Nadkarni 2000, Holz & Gradstein 2005). In general, growth of epiphytes seems to be very slow and this is a major hindrance to their speedy recovery following deforestation (Zotz 1995).

We also found that epiphytes may be good indicators of disturbance, especially bryophytes (Nöske 2005). About 40% of the bryophyte species occurred with high fidelity in only one of the three habitat types studied as compared with ca. 10% of lichens. Comparison of species numbers, furthermore, revealed species ratio's that correlated significantly with rate of disturbance. A two to one ratio of lichen to bryophyte species was characteristic for forest canopies and isolated trees; in the forest understory, however, the ratio was reversed. Thus, the lichen / bryophyte ratio may be used as a powerful indicator of microclimatic conditions and degree of human influence in tropical montane forests.

Conclusions. Our studies on the impact of forest disturbance on epiphyte diversity revealed the paramount importance of the microclimate and the closure of the forest canopy. Disruption of the canopy leads to microclimatic changes which in turn affect the epiphyte communities. In addition, changes in tree species composition and host tree characteristics play an important role. Species losses following disturbance vary considerably among the different groups of epiphytes and are most severe among orchids, bromeliads and filmy ferns. Moreover, the drought-intolerant shade epiphytes of the forest understory are more strongly impacted than the drought-tolerant species of the forest canopy. Species turnover along the disturbance gradient is high and recovery of the epiphyte communities in the regrowing forest is very slow. Thus, the epiphyte flora of 50 years old secondary forest is still very different from that of the natural forest.

We also found a preliminary explanation for the contrasting figures in the literature concerning the impact of deforestation on the diversity of vascular epiphytes (Werner et al. 2005). A comparison of the macroclimates at the different study sites showed that areas where no changes after disturbance occurred had a permanently wet or dry climate. Air humidity regimes inside and outside the forest were rather similar here throughout the year and opening up of the forest canopy had apparently little effect on the microclimate, and hence on vascular epiphyte diversity (bryophyte diversity may be impacted, however; Werner & Gradstein, in press). Areas where strong decline of species richness occurred, however, had a moderately seasonal, humid climate (1-3 month dry season). Here, air humidity regimes inside and outside the forest were clearly different and opening-up of the forest canopy resulted in significant humidity changes, which in turn affected the epiphytes. Based on these observations we concluded that major losses of vascular epiphyte diversity due to forest disturbance mainly occur in humid areas with a moderately seasonal climate.

Impact of global warming on epiphytes

Evidence is arising that organisms are affected in various ways by the current climatic changes. In a recent review on biodiversity and climate change, Lovejoy & Hannah (2005) formulated that global warming may impact biological organisms in four different ways: (1) by changes in the local abundance of species, (2) by changes in community structure, (3) habitat shifts, with species moving towards habitats with cooler microclimates, and (4) range shifts, with species moving towards higher latitudes or elevations. Most of the evidence, however, is derived from modelling work or from observations on the behaviour of animal taxa (birds, butterflies, amphibia, etc.). Empirical studies on the impact of climate change on plants are still scarce and only very few deal with epiphytes, in spite of their predicted sensitivity to global warming (Lugo & Scatena 1992, Benzing 1998, Gignac 2001).

First empirical data on the impact of global warming on epiphytes are arising from work on bryophytes and lichens in Europe. Based on a study of herbarium records, Frahm & Klaus (1997, 2001) found that 32 subtropical species had extended their ranges several hundred kilometres east- and north-eastwards into Central Europe in recent decades. These range extensions correlated with a recent increase of the mean winter temperature in the area by ca. 1.5 degree Celsius. In The Netherlands, Herk et al. (2002) observed significant shifts in the geographical ranges of epiphytic lichens based on a monitoring of the lichen flora in permanent plots during almost 25 years. During the first fifteen years, significant floristic changes occurred which were could be explained by changes in the pollution levels of the air. In the 1990s, however, when air pollution was no longer a problem in the country, significant changes occurred which correlated with the geographical distribution of the species. Thus, they observed a significant increase in the frequency of warm-temperate species and a significant decrease of cold-temperate ones. Moreover, several tropical species were newly detected in the country. The recent expansion of tropical and warm-temperate bryophytes and lichens in

Europe is striking first evidence of the possible impact of global warming on the epiphytic flora of the temperate regions of the Northern Hemisphere.

In the tropics, an assessment of the possible impact of global warming on epiphytes is hampered by lack of long-term distribution records and monitoring. Nevertheless, first evidence is arising from transplantation experiments. Nadkarni & Solano (2002) studied the possible response of vascular epiphytes to simulated climate change in Costa Rica. They translocated four common species of epiphytes, a tank bromeliad (*Guzmania pachystylis*), a woody shrub (*Clusia* sp.), a scandent herb (*Peperomia* sp.) and a fern (*Elaphoglossum* sp.) from cloud forest to drier climatic conditions with less cloud water ca. 150 m downslope. After one year, the translocated plants had significant leaf mortality and reduced production of new leaves; control plants, however, showed no negative effects (Fig. 6). Moreover, plants were more strongly affected when translocated during the dry season than during the wet season. Unexpectedly, the authors also observed the invasion of terrestrial species into the canopy community after the death of the transplants. The results indicated that climate change may have rapid negative effects on the productivity and longevity of epiphyte species and may lead to compositional changes of the epiphytic community.

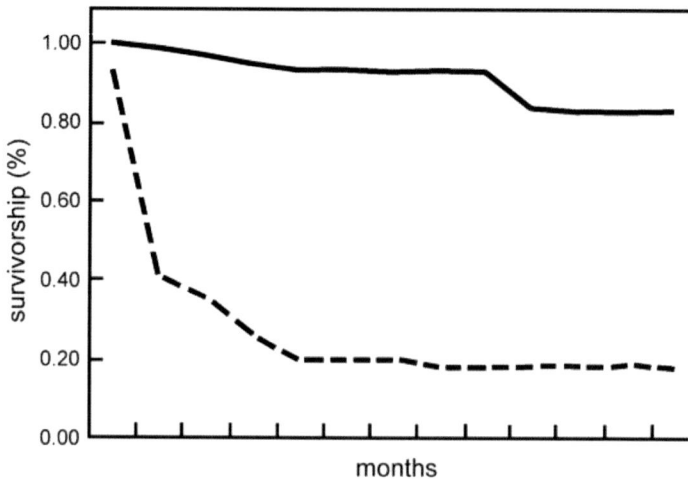

Figure 6. Survival of four species of epiphytes after translocation from montane cloud forest to warmer and drier climatic conditions (from 1480 to 1330 m) in Monteverde, Costa Rica. After one year, about 80% of the translocated plants had died. Broken line: translocated plants (transplanation during the dry season). Straight line: control plants. (after Nadkarni & Solano 2002).

In the high Andes of Bolivia, Jacome et al. (in prep.) carried out a translocation experiment on whole epiphytic communities. Branches of one meter length covered by dense mats of epiphytic bryophytes were cut and moved to a drier climate downslope, from 3000 m to 2700 and 2500 m respectively (twenty branches in each elevation). At each elevation, the cut branches were hung from trees in such a manner that the original growth conditions (light, exposition) were restored. After two years, considerable changes in the structure of the communities and the abundances of individual species were noted. Some species had clearly increased in abundance after 2 year period, at the cost of others which had become scarcer. The observed changes were gradual ones and were most pronounced at 2500 m, where temperatures were 2.5 degree Celsius higher than at the control site.

The translocation experiments in Costa Rica and Bolivia clearly show that global warming may impact the growth and community structure of the tropical epiphytes. The changes are gradual ones and the impact differs among species. The results are suggestive of significant changes in the composition of the rich epiphyte flora of the tropical montane forest over time due to climate change. Further monitoring over longer periods of time is necessary to demonstrate the predicted long-term effects of global warming on tropical epiphytes.

Acknowledgements. I thank Nalini Nadkarni, Jorge Jacome and Michael Kessler for data on epiphyte ecology and climate change, Lou Jost and Florian Werner for photographs, and Michael Kessler and Nalini Nadkarni for comments on the manuscript.

References

Acebey C, Gradstein SR, Krömer T (2003) Species richness and habitat diversification of bryophytes in submontane rain forest and fallows of Bolivia. Journal of Tropical Ecology 19: 9-18

Barthlott W, Schmitt-Neuerburg V, Nieder J, Engwald S (2001) Diversity and abundance of vascular epiphytes: a comparison of secondary vegetation and primary montane rain forest in the Venezuelan Andes. Plant Ecology 152: 145-156

Benzing DH (1998) Vulnerabilities of tropical forests to climate change: the significance of resident epiphytes. Climate Change 39: 519-540

Bermudes D, Benzing DH (1989) Fungi in neotropical epiphyte roots. Biosystems 23: 65-73

Bernal R, Celis M, Gradstein SR (2007) Plant diversity of Colombia catalogued. Taxon 56: 273

Cornelissen JHC, Gradstein SR (1990) On the occurrence of bryophytes and macroclichens in different lowland rain forest types at Mabura Hill Guyana. Tropical Bryology 3: 29-35

Dunn RR (2000) Bromeliad communities in isolated trees and three successional stages of an Andean cloud forest in Ecuador. Selbyana 21: 137-143

Ellwood MD & Foster W (2004) Doubling the estimate of invertebrate biomass in a rainforest canopy. Nature 429: 549-551

Frahm JP (2003) Climatic habitat differences of epiphytic lichens and bryophytes. Cryptogamie, Bryologie 24: 3-14

Frahm JP, Klaus D (1997) Moose als Indikatoren von Klimafluktuationen in Mitteleuropa. Erdkunde 51: 181-190

Frahm JP, Klaus D (2001) Bryophytes as indicators of recent climate fluctuations in Central Europe. Lindbergia 26: 97-104

Gignac LD (2001) Bryophytes as indicators of climate change. The Bryologist 104: 410-420

Gradstein SR, Nadkarni NM, Krömer T, Holz I, Nöske N (2003) A protocol for rapid and representative sampling of epiphyte diversity of tropical rain forests. Selbyana 24: 87-93

Gradstein SR (2006) The lowland cloud forest of French Guiana – a liverwort hotspot. Cryptogamie, Bryologie 27: 141-152

Gradstein SR (in press) The tropical lowland cloud forest – a neglected forest type. In: Bruijnzeel LE, Juvik JA (eds) Mountains in the mist. Hawaii University Press, Hawaii

Herk CM van, Aptroot A, Dobben HF van (2002) Long-term monitoring in the Netherlands suggests that lichens respond to global warming. The Lichenologist 34: 141-154

Hietz-Seifert U, Hietz P, Guevara S (1996) Epiphyte vegetation and diversity on remnant trees after forest clearance in southern Veracruz, Mexico. Biological Conservation 75: 103-111

Hilt N, Fieldler K (2005) Diversity and composition of Arctiidae moth ensembles along a succession gradient in the Ecuadorian Andes. Diversity and Distributions 11: 387-398

Hilt N, Brehm G, Fiedler K (2006) Diversity and ensemble composition of geometrid moths along a successional gradient in the Ecuadorian Andes. Journal of Tropical Ecology 22: 155-166

Hofstede RGM, Wolf JHD, Benzing DH (1993) Epiphytic biomass and nutrient status of a Colombian upper montane rain forest. Selbyana 14: 37-45

Holz I, Gradstein SR (2005) Cryptogamic epiphytes in primary and recovering upper montane oak forests of Costa Rica - species richness, community composition and ecology. Plant Ecology 178: 89-109

Ibisch P (1996) Neotropische Epiphytendiversität - das Beispiel Bolivien. Galunder Verlag, Wiehl

Krömer T, Gradstein SR (2004) Species richness of vascular epiphytes in two primary forests and fallows in the Bolivian Andes. Selbyana 24: 190-195

Liede-Schumann S, Breckle S (eds) (2008) Provisional Checklists of Flora and Fauna of the San Francisco valley and its surroundings, Southern Ecuador. Ecotropical Monographs 4

Lovejoy TE, Hannah L (2005) Climate change and biodiversity. Yale University Press, New Haven

Lugo AE, Scatena F (1992) Epiphytes and climate change research in the Caribbean: a proposal. Selbyana 13: 123-130

Nadarajah P, Nawami A (1993) Mycorrhizal status of epiphytes in Malaysian oil palm plantations. Mycorrhiza 4: 21-25

Nadkarni NM (1984) Epiphyte mats and nutrient capital of a neotropical elfin forest. Biotropica 16: 249-256

Nadkarni NM (2000) Colonization of stripped branch surfaces by epiphytes in a lower montane cloud forest. Biotropica 32: 358-363

Nadkarni NM, Longino JT (1990) Invertebrates in canopy and ground organic matter in a Neotropical montane forest, Costa Rica. Biotropica 22: 286-289

Nadkarni NM, Matelson TJ (1989) Bird use of epiphytic resources in neotropical trees. Condor 91: 891-907

Nadkarni NM, Solano R (2002) Potential effects of climate change on canopy communities in a tropical cloud forest: an experimental approach. Ecologia 131: 580-594

Nöske N (2005) Effekte anthropogener Störung auf die Diversität kryptogamischer Epiphyten (Flechten, Moose) in einem Bergregenwald in Südecuador. Dissertation, Universität Göttingen

Nöske N, Hilt N, Werner F, Brehm G, Fiedler K, Sipman HJ, Gradstein SR (in press) Disturbance effects on diversity in montane forest of Ecuador: sessile epiphytes versus mobile moths. Basic and Applied Ecology 9: 4-12

Turner IM, Tan HTW, Wee YC, Ibrahim AB, Chew PT, Corlett RT (1994) A study of plant species extinction in Singapore: lessons for the conservation of tropical biodiversity. Conservation Biology 8: 705-712

Werner F, Homeier J, Gradstein SR (2005) Diversity of vascular epiphytes on isolated remnant trees in the montane forest belt of southern Ecuador. Ecotropica 11: 21-40

Werner F, Gradstein SR (in press) Diversity of dry forest epiphytes across a gradient of human disturbance in the tropical Andes. Journal of Vegetation Science

Williams-Linera,G (1990) Vegetation structure and environmental conditions of forest edges in Panama. Journal of Ecology 78: 356-373

Yanoviak SP, Nadkarni NM, Solano R (2007) Arthropod assemblages in epiphyte mats of Cost Rican cloud forest. Biotropica 36: 202-210

Zotz G (1995) How fast does an epiphyte grow? Selbyana 16: 150-154

Biodiversity and Ecology Series (2008) 2: 67-78
The Tropical Mountain Forest – Patterns and Processes in a Biodiversity Hotspot
edited by S.R. Gradstein, J. Homeier and D. Gansert
Göttingen Centre for Biodiversity and Ecology

Mycorrhizal fungi and plant diversity in tropical mountain rainforest of southern Ecuador

Ingrid Kottke[1], Adela Beck[1], Ingeborg Haug[1], Sabrina Setaro[1] and Juan Pablo Suarez[2]

[1]Institute of Botany, Department of Special Botany and Mycology, Eberhard-Karls-University Tübingen, Auf der Morgenstelle 1, 72076 Tübingen, Germany, Ingrid.Kottke@uni-tuebingen.de
[2] Centro de Biología Celular y Molecular, Universidad Técnica Particular de Loja, San Cayetano Alto s/n C.P. 11 01 608, Loja, Ecuador

Abstract. Mycorrhizal fungi, because of their obligate symbiotic interaction with plant roots, may either promote or restrict plant diversity depending on broad or narrow plant-fungus relationships. Inventories based on morphotyping and DNA sequencing were carried out on the mycorrhizal fungi associated with 115 species of trees (40 families), 20 of ericads and 4 of epiphytic orchids in a tropical mountain rainforest area of about twelve ha in southern Ecuador. Results indicated that diverse Glomeromycota with broad host ranges may promote high tree diversity while diverse but plant-family restricted Sebacinales likely support closely related Andean ericads. Similarly, distinct Tulasnellales and Sebacinales are associated with closely related species of epiphytic orchids. Ectomycorrhizal fungi were specifically associated with three Nyctaginacean trees and with one member of Melastomataceae. We conclude that the extraordinary high plant diversity of the tropical Andean forest is predominantly promoted by a broad range of mycorrhizal fungi but selected trees are supported by specific fungi.

Introduction

The tropical Andes constitute a hotspot of plant diversity (Myers et al. 2000) and this applies also to the mountain rainforest of the Reserva Biológica San Francisco (RBSF) in South Ecuador, situated on the eastern slope of the Cordillera del Consuelo (Beck & Kottke 2008, see also chapter 13 of this volume). This forest is especially rich in tree species and families, ericads and orchids (Homeier & Werner 2008, Homeier et al. 2008). Members of these plant groups are known to depend on mycorrhizal fungi for seedling establishment and survival in nutrient-limited environments (Smith & Read 1997). However, no information was available on their mycorrhizal state and involved fungi in the tropical mountain rainforest. A first inventory of the mycorrhizal state of a large number of trees, terrestrial and hemiepiphytic ericads, as well as some epiphytic orchids in the RBSF revealed persistently well developed mycorrhizas (Kottke et al. 2004, Setaro et al. 2006a, Suarez et al. 2006). Morpho-typing and sequencing of the mycorrhizal fungi directly from the mycorrhizas revealed distinct associations for trees (mainly arbuscular mycorrhizas/AM, rarely ectomycorrhizas/ECM), ericads (cavendishioid mycorrhizas/CM for Andean ericads, ericoid mycorrhizas/ERM for non-Andean ericads) and pleurothallid orchids (orchid mycorrhizas/OM) (Kottke et al. 2008a). The detected fungi comprised Glomeromycota (AM), Ascomycota (ECM, ERM) and Basidiomycota (ECM, CM, OM) (Kottke et al. 2008b).

Experiments have shown that mycorrhizal fungi influence plant diversity (Grime et al. 1987, van der Heijden et al. 1998). Broad host range and multiple numbers of mycorrhizal fungi promote plant diversity, while narrow host range of few, specifically efficient fungi supports a low number of plant species resulting in monodominant forests (Kottke 2002) or specific myco-heterotrophic associations with restricted numbers of plant species (Taylor et al. 2002). The previously obtained results came from investigations of grassland communities and temperate forests. Tropical mountain rainforest are not only much more diverse but in addition show peculiarities not found in temperate areas. First, most of the tree species occur widely scattered, often only once in an area of 10,000 ha; only very few tree species are numerous and occasionally dominant (Homeier 2004). Second, closely related plant species theoretically expected to outcompete each other may occur abundantly within small areas. The latter phenomenon applies especially to the hemiepiphytic ericads and the epiphytic pleurothallid orchids. In this paper we discuss the results of our mycorrhiza study in tropical mountain rainforest with reference to four hypotheses:

1. Numerous and redundant mycobionts will promote plant diversity.
2. Specifically associated mycobionts will only support the frequency of the host species.
3. Mycobionts restricted to a plant family or genus but therein redundant will promote co-occurrence of closely related species.
4. Mycorrhizas are rare in tropical mountain forest and therefore do not influence plant diversity.

Numerous arbuscular mycorrhizal fungi with a broad host range are associated with tropical mountain rainforest trees

Root samples of 115 tree species were collected in 15 plots along an elevational gradient between 1850 and 2300 m (Homeier 2004). The mycorrhizal state of the roots was investigated by light microscopy after staining with methyl blue (Fig. 1).

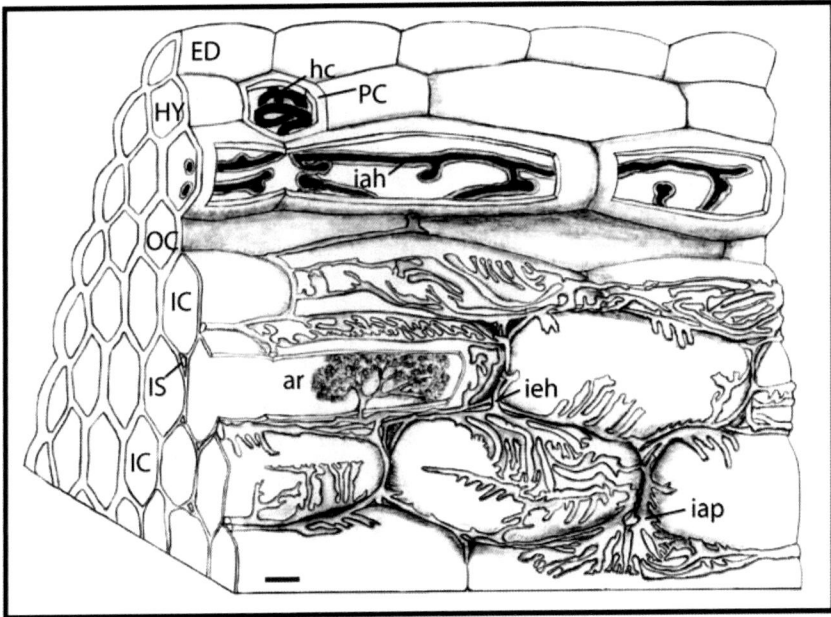

Figure 1. Morphotype of arbuscular mycorrhiza associated with *Alzatea verticillata*. Note hyphal coils (hc) in the hypodermis passage cells (HY, PC) and outer cortex (OC; iah intracellular hyphae), fine, finger-like branched hyphae (iap, ieh) in the intercellular space (IS) between the inner cortical cells (IC) and intracellular arbuscules (ar); bar 10 μm (Beck et al. 2005).

Associated fungi were identified by DNA sequencing. In total, 112 tree species belonging to 40 different families were found to form arbuscular mycorrhizas (Kottke & Haug 2004). Fourteen morphotypes of arbuscular mycorrhizas corresponding to the three main groups of Glomeromycota (Glomeraceae, Gigasporaceae Acaulosporaceae) were detected on one single tree species, *Alzatea verticillata* (Alzateaceae; Beck et al. 2007). The unexpectedly high fungal diversity of Glomeromycota was supported by phylogenic analysis of nucSSU sequences (Beck et al. 2007).

Glomeraceae, Acaulosporaceae, Gigasporaceae, Paraglomeraceae and Archaeosporaceae were found by DNA sequencing of the mycobionts of 40 further tree species belonging to 21 families. Members of *Glomus* group A were the most frequent mycobionts and molecular phylogeny resolved 24 different sequence types, each containing sequences with < 1% bp differences (Kottke et al. 2008b). Fungal sequences obtained from mycorrhizas of different forest tree species clustered within many of the sequence types (Fig. 2, marked by F). Assuming that the sequence types as defined here correspond to fungal species, our results indicate a frequent and broad sharing of mycobionts by the tropical trees. We conclude that

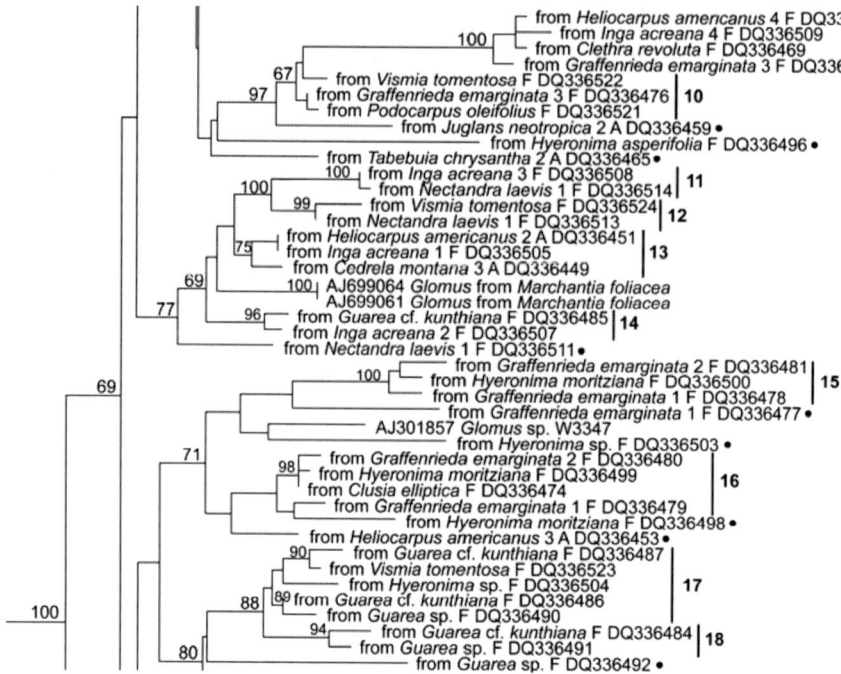

Figure 2. Glomeralean nucSSU sequence types on trees in the investigated tropical mountain forest; relevant sequence types are bold-faced numbers 9-12 and 14-18. Sequence types are defined by < 1% bp differences (for complete tree see Kottke et al. 2008b).

our first hypothesis holds true for this tropical mountain rainforest and that numerous and redundant AM mycobionts promote the high tree diversity. This result is in agreement with findings from grassland communities (Johnson et al. 2005).

Ectomycorrhiza-forming fungi are rare and have narrow host ranges in the tropical mountain rainforest

Only four tree species were found to form ectomycorrhizas. Three species (two of *Neea* and one of *Guapira*) belong to Nyctaginaceae, a family known to form ectomycorrhizas in tropical forests of South America (Alexander & Högberg 1986, Moyersoen 1993). *Graffenrieda emarginata*, however, was the first member of Melastomataceae shown to form ECM and simultaneously also AM (Haug et al. 2004). Sections through the mycorrhizas revealed Glomeromycota in the inner cortical layer, and an Ascomycete forming a thin hyphal sheath (Fig. 3) and a superficial Hartig net. DNA sequences of the Ascomycete clustered within the *Rhizoscyphus* (=*Hymenoscyphus*) *ericae* aggr. (Haug et al. 2004), a group of fungi forming ericoid

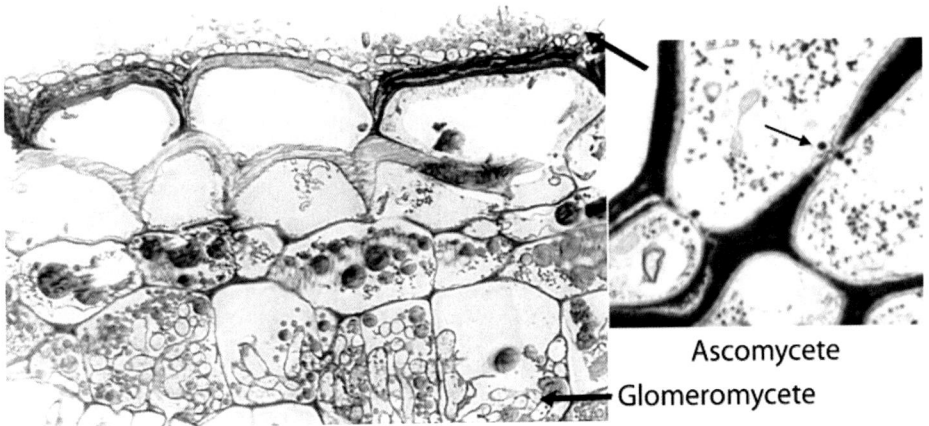

Figure 3. *Left*: micrograph of a section through the mycorrhizas of *Graffenrieda emarginata* revealing a superficial hyphal sheath formed by an Ascomycete and intracellular hyphae of a Glomeromycete in the inner cortical layer. *Right*: enlargement of the simple pore and the Woronin bodies (arrow) that characterize Ascomycetes (Haug et al. 2004).

mycorrhizas and shown to mobilize Norg and Fe at limiting soil conditions (Read & Perez-Moreno 2003). Interestingly, *G. emarginata* is a frequent tree in the RBSF, even dominating at 2000-2200 m (Homeier 2004). Although experimental proof is lacking, we speculate that *G. emarginata* is supported by this Ascomycete in a similar way as shown for the ericads. The specifically associated mycobiont, then, would support just this one species, as corresponding to our second hypothesis.

Nyctaginaceaen trees provide other examples of specifically associated ECM fungi in the RBSF area (Fig. 4; Haug et al. 2005). *Neea* sp. 1 is a hazelnut-bush-like tree (Fig. 4a) occurring along the steep slopes of the San Francisco river and on old road sides in the RBSF. It was found associated with *Russula puigarrii* as identified by DNA sequences from mycorrhizal hyphae (Fig. 4b) and fruiting bodies (Fig. 4c). Additionally, mycorrhizas were detected by morpho-typing and DNA-sequencing of two Thelephoracean species (Fig. 4d), one species of *Lactarius* (Fig. 4e) and an Ascomycete (not shown). None of these fungal species was found associated with other tree species in the study area. Apparently the ECM mycobionts support only *Neea* sp. 1, at the harshest conditions at the forest border. Two further Nyctaginacean species, *Guapira* sp. and *Neea* sp. 2, occur scattered in steep ravines in closed forest and display a very different ECM state. ECMs are formed highly localized on long roots with persisting root hairs by a species of Thelephoraceae, a different one for each of the two tree species (Haug et al. 2005). Occasionally, hyphae of Glomeromycota were found in the inner cortex, indicating a switch from AM to ECM, but without obvious ecological consequences. Influence of mycorrhizas on frequency of these two species is not evident, thus supporting our fourth hypothesis.

Figure 4. Ectomycorrhiza forming *Neea* sp. 1 (a), ectomycorrhiza (b) and fruiting body of *Russula puigarrii* (c), mycorrhiza formed by a Thelephoraceae (d), and *Lactarius* sp. (e).

Mycorrhiza-forming fungi of hemi-epiphytic ericads and epiphytic orchids are heterobasidiomycetes and redundantly associated within each plant family

Species of the Andean hemi-epiphytic ericad clade with hummingbird pollinated flowers (Fig. 5a, b) are mostly endemic to the northern Andes (Powell & Kron 2003). Epiphytic orchids occur abundantly in the region and are extremely species-rich, especially in the subfamily Pleurothallidinae (Fig. 5). Roots were collected from 20 species of ericads, 15 belonging to the Andean clade and 5 to the bushy, non-Andean clade, and from 4 species of pleurothallid orchids. Sampling was done within and outside the tree plots along the altitudinal gradient described above. Transmission electron microscopy revealed heterobasidiomycetes (Sebacinales, Tulasnellales, Ceratobasidiales) as the only fungi of orchid mycorrhizas and Sebacinales as the dominating fungi of the Andean ericad clade, the latter forming a previously unknown ectendomycorrhiza (Fig. 6; Setaro et al. 2006a, 2006b).

Sequences of the associated fungi were obtained by use of specific primers for nucLSU of Sebacinales and Tulasnellales (Setaro et al. 2006b, Suarez et al. 2006). Diverse Sebacinales of Group B were found in mycorrhizas of ericads and orchids, but molecular phylogenetic analysis resolved the sequences of ericad and orchid mycobionts in different clades. However, identical or closely related sequences of

Figure 5. a, flower of *Cavendishia nobilis*, an Andean clade, hummingbird pollinated ericad. **b,** hemi-epiphytic growth of *Cavendishia nobilis* in tropical mountain forest. **c,** flowering *Stelis superbiens*. **d,** mycorrhizal roots of *Pleurothallis* sp. on tree trunk.

Sebacinales obtained from diverse ericad species occurred within the same individual clades (Kottke et al. 2008). Tulasnellales were the most frequent fungi in the epiphytic orchids (13 sequence types as defined by < 1% bp difference in nucLSU sequences). Sequences of mycobionts from different orchid species were found within the individual sequence types (Fig. 7; Suarez et al. 2006). The results obtained for ericads and pleurothallid orchids indicate that the mycobionts are restricted to the respective plant families, but therein redundant. Mycobionts will,

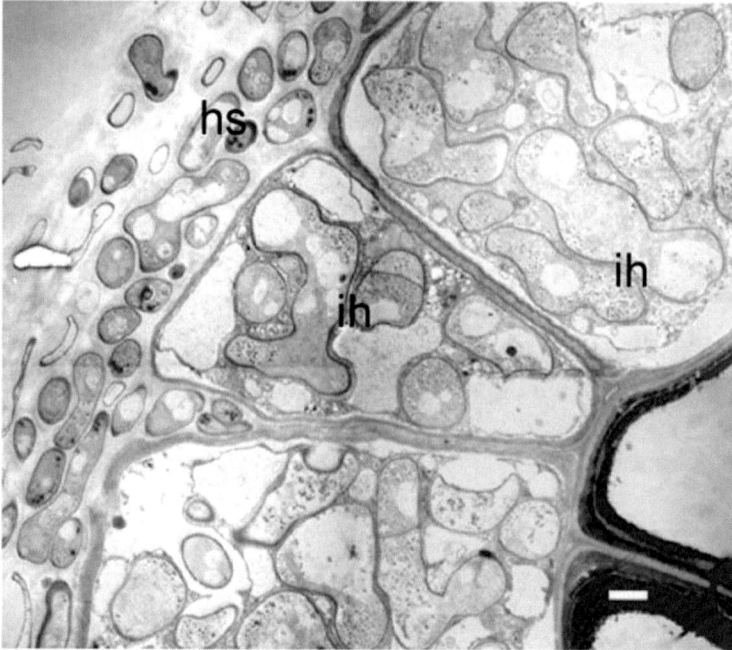

Figure 6. Transmission electron micrograph of a section through the mycorrhiza of *Psammisia guianenesis* (Andean clade ericads). Note the hyphal sheath (hs), the intercellular hyphae reminiscent of a Hartig net (hn) and the intracellular colonization of the cortex by blown up, septate hyphae (ih) (Setaro et al. 2006a).

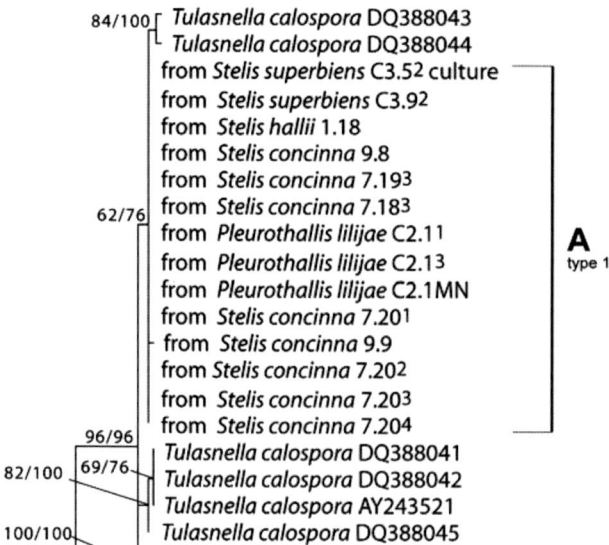

Figure 7. *Tulasnella* sequences differing < 1% (A type 1) are shared by several orchid species (Suarez et al. 2006).

therefore, not restrict but support co-occurrence of closely related plant species, as proposed in our third hypothesis.

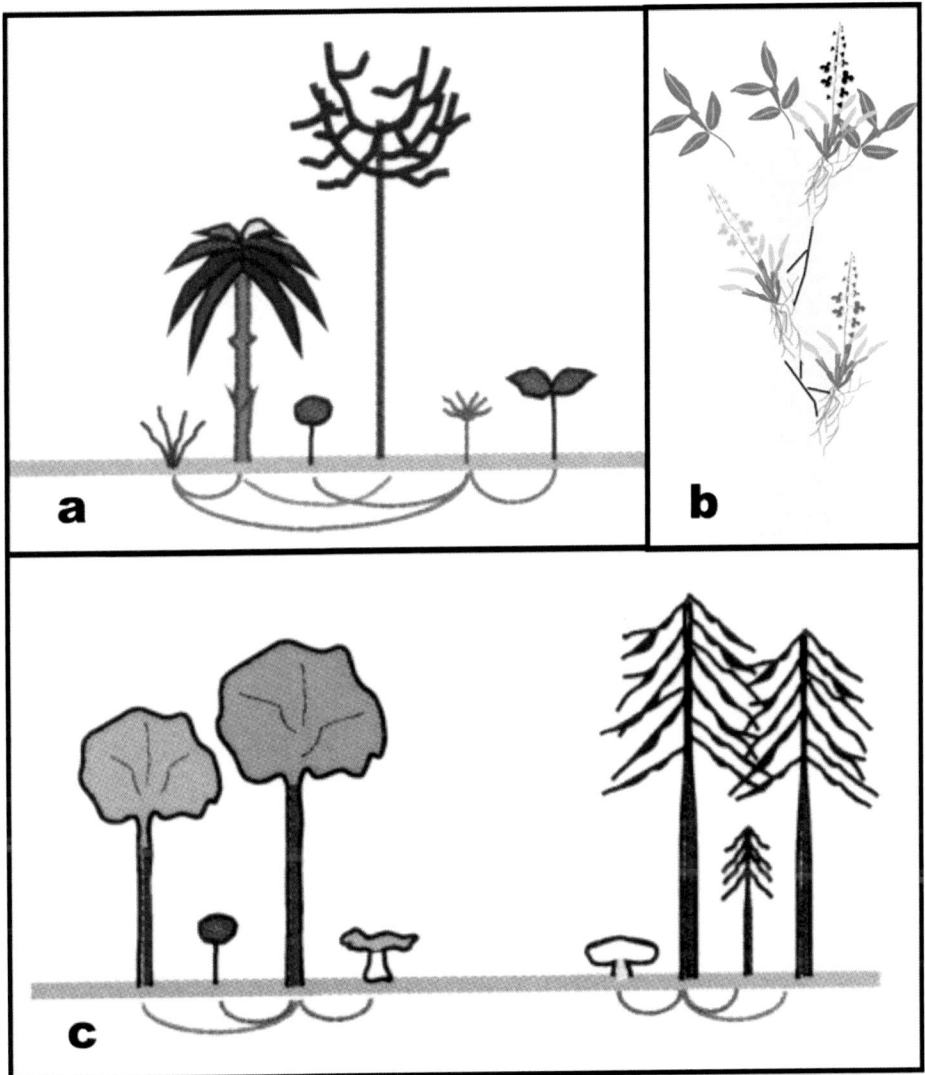

Figure 8. Influence of mycorrhizal associations on plant diversity. **a,** Glomeromycota with broad host rage promote plant diversity by supporting the off-spring of different host species. **b,** unspecific orchid mycorrhizal fungi support co-occurrence of closely related species. **c,** specifically associated ectomycorrhizal fungi only promote the off-spring of the host species, thus restricting tree diversity.

Conclusions

Our mycorrhizal studies in tropical mountain rainforest revealed different strategies of the fungal associations with respect to plant diversity (Fig. 8). The detected broad host ranges of AM fungi and specific associations of ectomycorrhizal fungi are in agreement with earlier findings in temperate regions (Hoeksema 1999). However, in contrast to temperate regions the number of AM fungi as determined by sequence types was unexpectedly high and that of ECM fungi very low. Thus, both AM fungal diversity and redundancy favour plant diversity by promoting offsprings independently from seed distance dispersal and scattered occurrence of parent trees (Fig. 8a). The low number of ECM fungi and their narrow host ranges can only support recruitment of the few ECM tree species (Fig. 8c). Redundancy of basidiomycete mycobionts in epiphytic orchids and Andean clade ericads was shown by us for the first time. Based on study of terrestrial taxa, orchids were considered to be linked to a narrow range of specific mycobionts (Shefferson et al. 2007) and ericads were mainly known to be associated with *Rhizoscyphus ericae* aggr. (Vrålstad et al. 2001). Assuming that the mycobiont sequence types as defined here correspond to species or closely related fungal taxa, our findings suggest that the co-occurrence of high numbers of closely related species of ericads and orchids in tropical mountain rainforest is supported by the sharing of mycorrhizal fungi.

Acknowledgements. We thank Jane Lempe, Andrea Abele, Vivian Jeske and Stefan Reidinger for valuable contributions, Jutta Bloschies for technical assistance, the members of the Department of Systematic Botany and Mycology, University of Tübingen, and the research group DFG FOR 402 for productive cooperation. The investigations were financially supported by the German Research Foundation (DFG FOR 402).

References

Alexander IJ, Högberg P (1986) Ectomycorrhizas of tropical angiospermous trees. New Phytologist 102: 541-549

Beck A, Kottke I, Oberwinkler F (2005) Two members of the Glomeromycota form distinct ectendomycorhizas with *Alzatea verticillata*, a prominent tree in the mountain rain forest of southern Ecuador. Mycological Progress 4: 11-22

Beck A, Haug I, Oberwinkler F, Kottke I (2007) Structural characterization and molecular identification of arbuscular mycorrhiza morphotypes of *Alzatea verticillata* (Alzateaceae), a prominent tree in the tropical mountain rain forest of South Ecuador. Mycorrhiza 17: 607-625

Beck E, Kottke I (2008) Facing a hotspot of biodiversity. Basic and Applied Ecology 9: 1-3

Grime JP, Mackey JM Hillier SH, Read DJ (1987) Floristic diversity in a model system using experimental microcosms. Nature 328: 420-422

Haug I, Lempe J, Homeier J , Weiß M, Setaro S, Oberwinkler F, Kottke I (2004) *Graffenrieda emarginata* (Melastomataceae) forms mycorrhizas with Glomeromycota and with a member of *Hymenoscyphus ericae* aggr. in the organic soil of a neotropical mountain rain forest. Canadian Journal of Botany 82: 340-356

Haug I, Weiß M, Homeier J, Oberwinkler F, Kottke I (2005) Russulaceae and Thelephoraceae form ectomycorrhizas with members of the Nyctaginaceae (Caryophyllales) in the tropical mountain rain forest of southern Ecuador. New Phytologist 165: 923-936

Hoeksema JD (1999) Investigating the disparity in host specificity between AM and EM fungi: lessons from theory and better-studied systems. Oikos 84: 327-332

Homeier J (2004) Baumdiversität, Waldstruktur und Wachstumsdynamik zweier tropischer Bergregenwälder in Ecuador und Costa Rica. Dissertationes Botanicae 391: 1-207

Homeier J, Werner FA (2008) Spermatophyta. In: Liede-Schumann S, Breckle SW (eds) Provisional Checklists of Flora and Fauna of the San Francisco valley and its surroundings, Southern Ecuador. Ecotropical Monographs 4

Homeier J, Werner FA, Breckle S, Gradstein SR, Richter M (2008) Potential vegetation and floristic composition of Andean forests in South Ecuador, with a focus on the RBSF. In: Beck E, Bendix J, Kottke I, Makeschin F, Mosandl R (eds) Gradients in a Tropical Mountain Ecosystem of Ecuador. Ecological Studies 198. Springer, Berlin, Heidelberg, New York, pp 87-100

Johnson D, IJdo M, Genney DR, Anderson IC, Alexander IJ (2005) How do plants regulate the function, community structure, and diversity of mycorrhizal fungi? Journal of Experimental Botany 56: 1751-1760

Kottke I (2002) Mycorrhizae - Rhizosphere determinants of plant communities. In: Waisel Y, Eshel A, Kafkafi U (eds) Plant Roots: The Hidden Half, 3rd ed. Marcel Dekker, Inc., New York, Basel, pp 919-932

Kottke I, Beck A, Oberwinkler F, Homeier J, Neill D (2004) Arbuscular endomycorrhizas are dominant in the organic soil of a neotropical montane cloud forest. Journal of Tropical Ecology 20: 125-129

Kottke I, Beck A, Haug I, Setaro S, Jeske V, Suárez JP, Pazmiño L, Preußing M, Nebel M, Oberwinkler F (2008a) Mycorrhizal state and new and special features of mycorrhizae of trees, ericads, orchids, ferns and liverworts. In: Beck E, Bendix J, Kottke I, Makeschin F, Mosandl R (eds) Gradients in a Tropical Mountain Ecosystem of Ecuador. Ecological Studies 198. Springer, Berlin, Heidelberg, New York, pp 137-148

Kottke I, Haug I (2004) The significance of mycorrhizal diversity of trees in the tropical mountain forest of southern Ecuador. Lyonia 7: 50-56

Kottke I, Haug I, Setaro S, Suárez JP, Weiß M, Preußing M, Nebel M, Oberwinkler F (2008b) Guilds of mycorrhizal fungi and their relation to trees, ericads, orchids and liverworts in a neotropical mountain rain forest. Basic and Applied Ecology 9: 13-23.

Moyersoen B (1993) Ectomicorrizas y micorrizas vesículo-arbusculares en Caatinga Amazonica del Sur de Venezuela. Scientia Guaianae 3: 1-183

Myers N, Mittermeier CG, da Fonseca GAB, Kent J (2000) Biodiversity hotspots for conservation priorities. Nature 403: 853-858

Powell EA, Kron KA 2003 Molecular systematics of the Northern Andean blueberries (Vaccinieae, Vaccinoideae, Ericaceae). International Journal of Plant Science 164: 987-995

Read DJ, Perez-Morena J (2003) Mycorrhizas and nutrient cycling in ecosystems - a journey towards relevance? New Phytologist 157: 475-492

Shefferson RP, Weiß M, Kull T, Taylor DL (2005) High specificity generally characterizes mycorrhizal association in rare lady's slipper orchids, genus Cypripedium. Molecular Ecology 14: 613-626

Setaro S, Oberwinkler F, Kottke I (2006a) Anatomy and ultrastructure of mycorrhizal associations of neotropical Ericaceae. Mycological Progress 5: 243-254

Setaro S, Weiß M, Oberwinkler F, Kottke, I. (2006b) Sebacinales form ectendomycorrhizas with *Cavendishia nobilis*, a member of the Andean clade of Ericaceae, in the mountain rain forest of southern Ecuador. New Phytologist: 169: 355-365.

Smith SE, Read DJ (1997) Mycorrhizal Symbiosis. 2nd ed. Academic Press, San Diego, London

Suárez JP, Weiß M, Abele A, Garnica S, Oberwinkler F, Kottke I (2006) Diverse tulasnelloid fungi form mycorrhizas with epiphytic orchids in an Andean cloud forest. Mycological Research 110: 1257-1270.

Taylor DL, Bruns TD, Leake JR, Read DJ (2002) Mycorrhizal specificity and function in myco-heterotrophic plants. Ecological Studies 157: 375-413

Van der Heijden MG, Klironomos JN, Ursic M, Moutoglis P, Streitwolf-Engel R, Boller T, Wiemken A, Sanders IR (1998) Mycorrhizal fungal diversity determines plant biodiversity, ecosystem variability and productivity. Nature 396: 69-72

Vrålstad T, Schumacher T, Taylor AF (2001) Mycorrhizal synthesis between fungal strains of the *Hymenoscyphus ericae* aggregate and potential ectomycorrhizal and ericoid hosts. New Phytologist 153: 143-152

Biodiversity and Ecology Series (2008) 2: 79-96
The Tropical Mountain Forest – Patterns and Processes in a Biodiversity Hotspot
edited by S.R. Gradstein, J. Homeier and D. Gansert
Göttingen Centre for Biodiversity and Ecology

The soil fauna of a tropical mountain rainforest in southern Ecuador: structure and functioning

Stefan Scheu, Jens Illig, Verena Eissfeller, Valentyna Krashevska, Dorothee Sandmann and Mark Maraun

Institute for Zoology, Darmstadt University of Technology, Schnittspahnstr. 3, 64287 Darmstadt, Germany, scheu@bio.tu-darmstadt.de

Abstract. The typical macro-decomposer species of lowland rain forests, such as termites, are rare or lacking in tropical mountain rainforests. We therefore hypothesized that small soil animals dominate the decomposer system and drive decomposition processes. To prove these hypotheses we investigated the density, diversity, community structure and functioning of soil microarthropods (oribatid mites) and soil microfauna (testate amoebae) at an altitudinal gradient of tropical mountain rainforests in southern Ecuador. Contrary to our expectations, at our study sites the density of microarthropods and microfauna is low and their importance during the initial stage of litter decomposition is limited. Unexpectedly and contrary to the thickness of organic layers, the density of soil microarthropods decreases with altitude. Consistent with the generally low density of decomposer animals, microbial biomass in litter and soil is also low. Presumably, mircoorganisms and decomposer animals suffer from high energy demand due to high temperatures with the acquisition of energy being limited by low food quality (strong sclerotization of leaves, low nutrient content of litter materials, high abundance of lignified roots). Contrary to superdiverse aboveground plant and animal groups, the species number of soil animals in the studied tropical mountain rain forests is similar to temperate forest ecosystems, suggesting that the latitudinal gradient in diversity of belowground biota is less pronounced than above the ground. In contrast to temperate and boreal forest systems where parthenogenetically reproducing soil animal species occur at high frequency, Oribatida in the studied tropical mountain rainforests were dominated by sexually reproducing species, in particular at higher elevation. This further suggests increasing nutrient limitation of decomposer biota with altitude.

Introduction

Tropical mountain rainforests are among the most species-rich ecosystems of the world (Myers et al. 2000). In these biodiversity 'hot spots' the number of species of animal and plant taxa may exceed that of temperate forests by orders of magnitude; e.g. the number of tree species per hectare in tropical mountain rainforests may be as high as 100, whereas in temperate forests only a handful of taxa are present per hectare (Whitmore 1998, Oosterhoorn & Kappelle 2000, Kessler et al. 2005). Similarly, the diversity of animal taxa, such as Coleoptera (Lucky et al. 2002) and Lepidoptera (Brehm & Fiedler 2003), is exceptionally high. Only few taxa, such as parasitic Hymenoptera, appear not to conform to the general rule of increasing diversity from temperate to tropical regions (Gauld et al. 1992). Compared to the few taxa above the ground, weak latitudinal gradients in species diversity appear to be common in soil animal taxa. By investigating in detail two of the most species-rich soil animal taxa, oribatid mites (Fig. 1) and testate amoebae (Fig. 2), we tested this assumption.

Compared to the aboveground system, animal communities below the ground generally received little attention and this is particularly true for tropical ecosystems (Anichkin et al. 2007). In tropical lowland ecosystems macrofauna decomposer animals including earthworms and termites form conspicuous components of the soil fauna. In contrast, in tropical mountain rainforests these animals are rare or lacking (Anichkin et al. 2007, Maraun et al. 2008). Parallel to temperate forest ecosystems where macrofauna is replaced by mesofauna at more acidic and nutrient poor sites (Schaefer 1991), the decomposer fauna of tropical mountain rainforests may be dominated by soil microarthropods and soil microfauna and consequently these groups may drive decomposition processes. Focussing on oribatid mites and testate amoebae we tested these hypotheses by investigating the decomposer community of tropical mountain rainforests along an altitudinal gradient in southern Ecuador. In addition to characterising the decomposer community structure and its changes with altitude we investigated structuring forces of soil animal communities and the role of soil fauna in decomposition processes.

Oribatid mites are small soil and litter colonizing arthropods reaching high density and diversity in virtually all ecosystems (Maraun & Scheu 2000). Commonly they are assumed to feed predominantly on fungi; however, it has been recently shown that trophically they are very diverse (Schneider et al. 2004), suggesting that trophic niche differentiation contributes significantly to species diversity. Testate amoebae are among the most abundant soil microfauna in particular at sites with low soil pH (Opravilova & Hajek 2006). However, despite their high density, their prominent position among soil microfauna and their presumed importance for decomposition processes, knowledge on testate amoebae of tropical forests is very limited.

A remarkable feature of soil animal taxa is the high frequency of parthenogenetic reproduction, which is most pronounced in oribatid mites (Maraun et al. 2003a, Domes et al. 2007). Interestingly, testate amoebae in general reproduce asexually; however, sexual processes may also occur (Valkanov 1962, Schönborn & Peschke 1988).

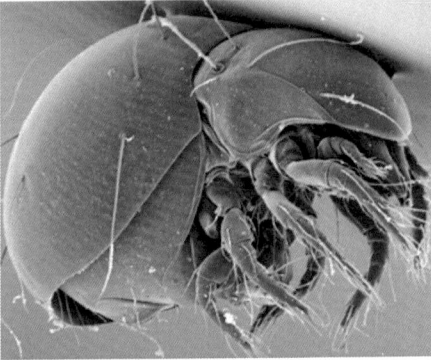
Mesoplophora cubana
Calugar & Vasiliu, 1977

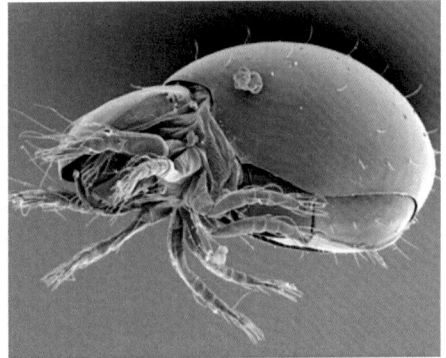
Protoplophora paraminisetosus
Niedbala & Illig, 2007

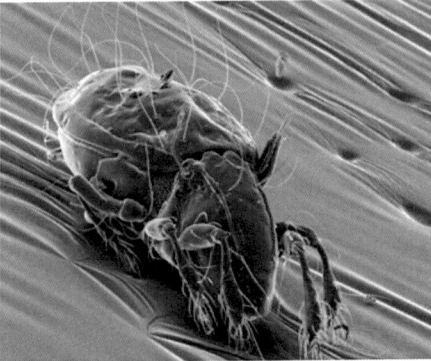
Dolicheremaeus bolivianus
Balogh & Mahunka, 1969

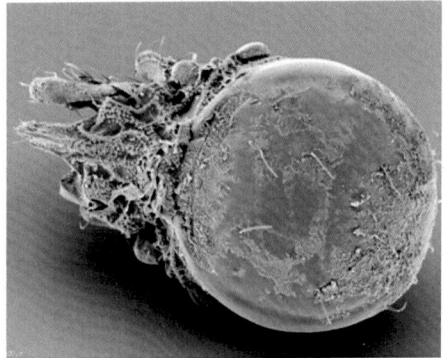
Rhynchoribates sp.

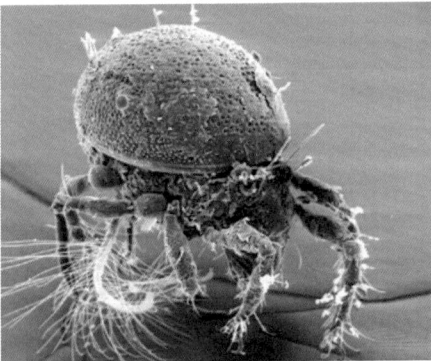
Hermannobates sp.

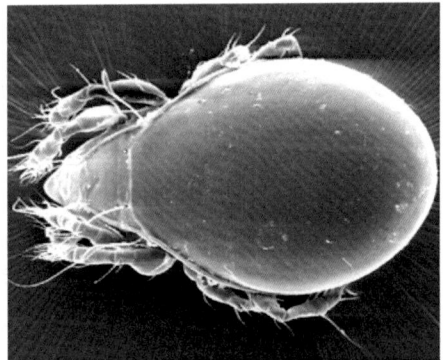
Unguizetes incertus
(Balogh & Mahunka, 1969)

Figure 1. Oribatid mites from tropical mountain rainforests of the region of the Reserva Biológica San Francisco in southern Ecuador.

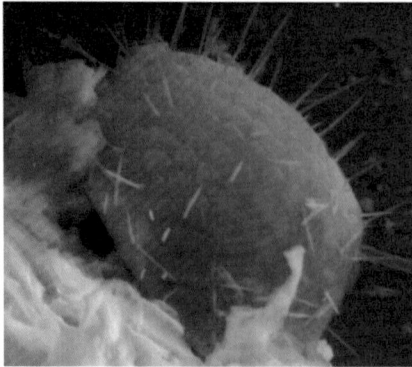

Placocista spinosa Carter, 1865

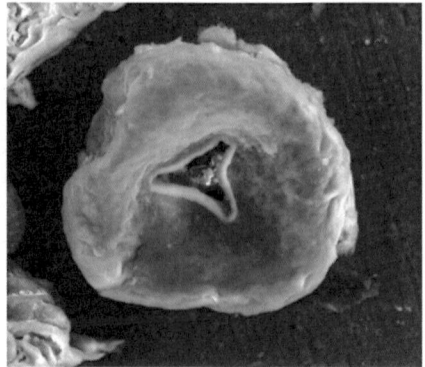

Trigonopyxis arcula Penard,1912

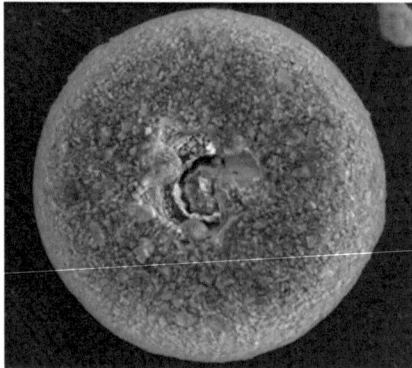

Lamtopyxis travei Bonnet, 1977

spec. nov.

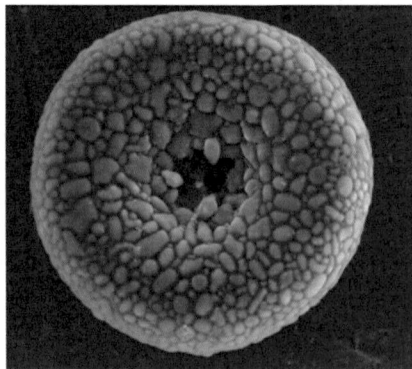

Cyclopyxis lithostoma Bonnet, 1974

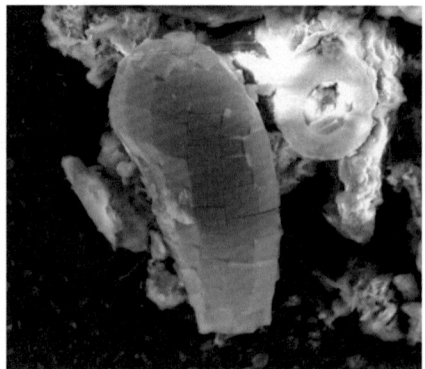

Quadrulella symmetrica var. *longicollis* Taranek, 1882

Figure 2. Testate amoebae from tropical mountain rainforests of the region of the Reserva Biológica San Francisco in southern Ecuador.

Frequent abandonment of males in soil animals raises the question on the function of sexual reproduction in general, and in particular why the necessity for sex might be relaxed in soil. Based on established theory, such as the Red Queen hypothesis (Hamilton 1980, Otto & Nuismer 2004), one may expect that soil animals are less exposed to attacks by parasites and pathogens. Further, the high incidence of parthenogenesis in soil might be related to difficulties in finding mating partners in the opaque and highly structured soil environment. Investigating the variability in the frequency of parthenogenetic reproduction in oribatid mites between very different localities, e.g. between temperate and tropical regions, may allow to prove the generality of the phenomenon of high incidence of pure female populations in soil. Further, it may allow identifying factors which may drive the abandonment of the production of males. To follow this line of arguments, we investigated the mode of reproduction in oribatid mites in the studied tropical rainforests.

Materials and methods

The study area is located in southern Ecuador within the Eastern Cordillera of the Andes in the province of Zamora-Chinchipe. Studies were carried out in the region of the Reserva Biológica San Francisco (RBSF, S 3°58', W 79°5') at the border of the Podocarpus National Park (Maraun et al. 2008). The sites are located on the steeply sloping (30°–50°), north-facing flank of the valley of the Rio San Francisco which drains in the Amazon basin. Along an elevation transect three study sites at 1000, 2000 and 3000 m a.s.l. were investigated. The maximum distance between the sites was 30 km. The study site at 1000 m was located in Bombuscaro near the province capital Zamora (S 04°06'54", W 78°58'02"), the one at 2000 m in the Reserva Biológica San Francisco (S 3°58'18", W 79°4'45") and the one at 3000 m in the Cajanuma area at the north-west gate of Podocarpus National Park south of Loja (S 04°06'711", W 79°10'581").

The sites are covered with mostly undisturbed mountain rainforest (Homeier et al. 2002). The soil types of the study sites are alumic acrisol at 1000 m, gley cambisol at 2000 m, and podzol at 3000 m (Soethe et al. 2006). The organic soil layer is thick and increases with elevation, the pH ranges between 3.0 and 5.5 (Wilcke et al. 2002) and is decreasing with elevation (Soethe et al. 2006). The bedrock consists mainly of weakly metamorphosed Paleozoic schists and sandstones with some quartz veins. The climate is semihumid with 8 to 10 humid months per year. Accordingly, annual rainfall is high with ca. 2200, 3500 and 4500 mm y^{-1} at 1000, 2000 and 3000 m, respectively. The mean annual air temperature is decreasing with altitude from 14.9, to 12.3 to 8.9°C at 1000, 2000 and 3000 m, respectively. The coldest month on average is August, the warmest November (Röderstein et al. 2005).

Density and community structure of oribatid mites and testate amoebae were investigated from soil samples taken with a corer (Ø 5 cm) and separated into litter layer and two deeper layers (0–3 cm, 3–6 cm depth). For extraction of oribatid mites the samples were transferred to the laboratory and extracted by heat (Kempson et al. 1963). For sexing oribatid mites were bleached with lactic acid as necessary and inspected under the microscope. Females were identified by their ovipositor and males by structures for spermatophore deposition. Litter and soil material for determination of testate amoebae was dried at room temperature and stored. For analysis the air dry litter and soil samples were rewetted for 24 h with sterile tap water (250 ml per 5 g litter or soil) to detach amoebal tests from the matrix. The next day, the samples were filtered through a 500 µm sieve to separate coarse organic particles. Testate amoebae were subsequently collected from the filtrate on a 250 µm mesh, and small forms were recovered by a final filter step using a 25 µm sieve. Microscopic slides were prepared and tests were identified and counted at 200x and 400x magnification.

Decompositon of leaf litter enclosed in litterbags of different mesh size was investigated along a smaller altitudinal gradient; two sites at 1850 (S 03°58'38", W 79°04'66") and 2280 m (S 03°58'96", W 79°04'41") were investigated (Illig et al. 2008). Litter of two of the dominating tree species at these two altitudes, *Graffenrieda emarginata* (1850 m) and *Purdiaea nutans* (2280 m), and the mixture of both was investigated. After drying at 65°C for 72 h, 10 g dry weight of the litter was filled in litterbags (20 x 20 cm) of different mesh-size (48 µm and 1 mm); the finer mesh-size prevents immigrating of soil microarthropods. A total of 144 litterbags were exposed in the field at 1850 and 2280 m. Litter mass loss and the colonisation of the litterbags by microarthropods were analysed after 2, 6 and 12 months.

Results and discussion

Density, diversity and reproductive mode of oribatid mites. The density of oribatid mites was at a maximum at 1000 m with 34,400 ind. m^{-2} (Fig. 3). This density is similar to that of other tropical forests (Plowman 1981), but much lower than that of acidic deciduous forests of the temperate zone where oribatid mites may reach densities up to 200,000 ind. m^{-2} (Maraun & Scheu 2000). Low density of oribatid mites (and other soil animal taxa; cf. Maraun et al. 2008) is rather surprising considering that the input of litter material in tropical forests considerably exceeds that of temperate and boreal forests. Hence, other factors than the amount of food resources must be responsible for low densities of soil animals.

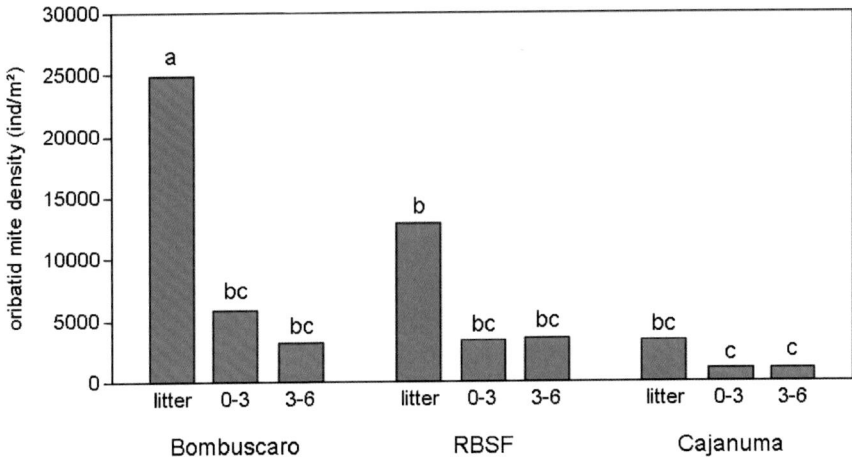

Figure 3. Oribatid mite densities at Bombuscaro (1000 m), RBSF (2000 m) and Cajanuma (3000 m) in the litter layer (litter) and two following deeper layers of a thickness of 3 cm (0–3 and 3–6). Bars sharing the same letter are not significantly different (Tukey's HSD test; P > 0.05).

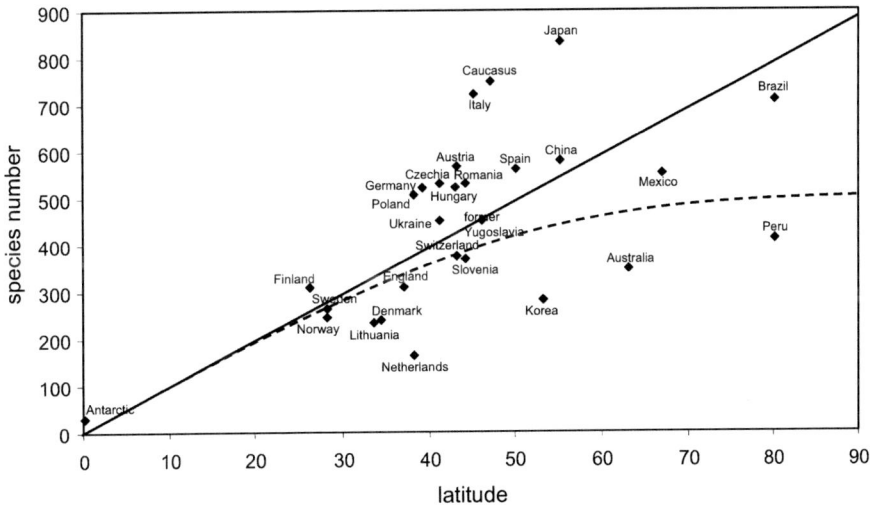

Figure 4. The species number - latitude relationship for oribatid mites of the world; solid line, upper limit of species numbers expected in tropical regions; dashed line, lower limit of species numbers expected in tropical regions (modified from Maraun et al., 2007).

Unexpectedly, and contrary to the thickness of organic layers, the density of oribatid mites decreased with altitude to 21,000 ind. m^{-2} at 2000 m and to only 5400 ind. m^{-2} at 3000 m a.s.l. This decline was particularly unexpected since the thickness of the organic layers increased with altitude and in temperate forests the density of microarthropods increases with the thickness of the organic layers, e.g. in moder as compared to mull systems (Maraun & Scheu 2000).

At the study sites 193 species/morphospecies of oribatid mites from 48 families were recorded, of which 30–40% might be new to science (Illig et al. 2007, Niedbala & Illig 2007a, b). Based on changes in oribatid mite species diversity with latitude (Maraun et al. 2007), a total of 500–800 species of oribatid mites may exist in Ecuador (Fig. 4). The observed species number therefore likely represents a large fraction of the existing species of the country, suggesting that species composition between sites (ß-diversity) changes little. High local species richness (α-diversity) and low ß-diversity is a characteristic feature of soil animal species in general (Schaefer 1999, Scheu 2005). Investigating local species richness in soil animals therefore contributes significantly to the knowledge of the total species diversity of the region. Interestingly, in other tropical mountain regions of America a similar number of species has been recorded (165 species in the Central American Cordillera de Talamanca; Schatz 2008) suggesting that the observed species number is characteristic for mountain rainforests. The recorded number of species of oribatid mites proves that oribatid mites are among the most diverse soil animal taxa. However, compared to the herbivore system above the ground the diversity of soil animals is low; e.g. the number of species of geometrid and arctiid moths is exceptionally high at the studied mountain rainforests (Brehm et al. 2005). This underlines the conclusion that the latitudinal gradient in species diversity in soil animal taxa is less steep than in plants and aboveground herbivore taxa (Maraun et al. 2007, 2008).

The community structure of oribatid mites in the studied mountain rainforests differs markedly from that in temperate forests. In particular derived groups of oribatid mites, such as Poronota and Pycnonotic Apheredermata, dominate indicating that these taxa are well adapted to the biotic and/or abiotic conditions in tropical mountain rainforests. Possibly, derived taxa of oribatid mites benefit from large and strongly sclerotised leaves abundant in tropical forest ecosystems since juveniles of these taxa in part live inside dead leaves (Hansen 1999).

Compared to temperate forests, the number of parthenogenetic taxa of oribatid mites at the studied tropical mountain rainforest is small. In temperate forests about 50% of the species and up to 90% of the individuals of oribatid mites reproduce via parthenogenesis (Maraun et al. 2003b), whereas at the 1000 m site of the studied tropical forests (Bombuscaro) only 23% of the species and about 35% of the individuals reproduce via parthenogenesis, with the frequency of parthenogenetic species decreasing strongly at higher altitude (Fig. 5; V. Eissfeller

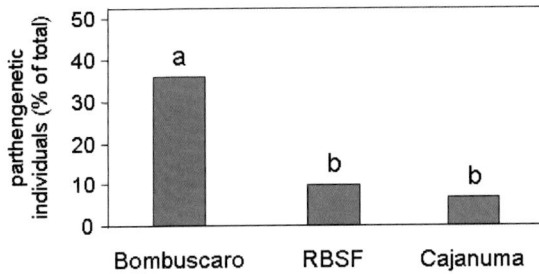

Figure 5. Percentage of parthenogenetic oribatid mite individuals in litter and soil at Bombuscaro (1000 m), RBSF (2000 m) and Cajanuma (3000 m). Bars sharing the same letter do not different significantly (Tukey's HSD test; P > 0.05; V. Eissfeller and M. Maraun, unpublished data).

and M. Maraun, unpublished data). According to the Red Queen hypothesis (Hamilton 1980, Otto & Nuismer 2004) this suggests increasing parasite/pathogen load towards the tropics and in the tropics with altitude. This, however, is unlikely to be true considering that the density of oribatid mites declines along these gradients and parasite load usually also declines with declining host density. Further, there is generally little evidence that oribatid mites are controlled by parasites/pathogens. Therefore, other factors appear to be responsible for the frequent abandonment of males in soil. Potentially, compared to temperate and boreal regions, soil animal species in tropical regions are confronted with a wider spectrum of resources and these resources are more strongly limiting soil animal populations. Theoretical considerations and modeling suggest that at these conditions parthenogenetic species are replaced by sexually reproducing species (Scheu & Drossel 2007). In fact, the low density but high species diversity of oribatid mites at the studied tropical mountain rainforests support this interpretation.

Density and diversity of testate amoebae. In contrast to oribatid mites the density of testate amoebae did not decrease with altitude but rather increased, however, not in a linear way. It was at a minimum at 1000 m, reached a maximum at 2000 m and was intermediate at 3000 m (Krashevska et al. 2007; Fig. 6). Density of testate amoebae was exceptionally low in the deeper layer at 1000 m, the only layer of mineral soil. Compared to temperate forests, even the high densities in the litter layer of 9,000–12,000 ind. g^{-1} dry wt as at 3000 and 2000 m are low; the density at 1000 m of ca. 3,000 ind. g^{-1} dry wt is exceptionally low, in particular considering the low pH. Generally, testate amoebae reach high densities in forest ecosystems of low soil pH (Opravilova & Hajek 2006). Surprisingly, variations in density of testate amoebae did not correlate closely with bacterial biomass, their

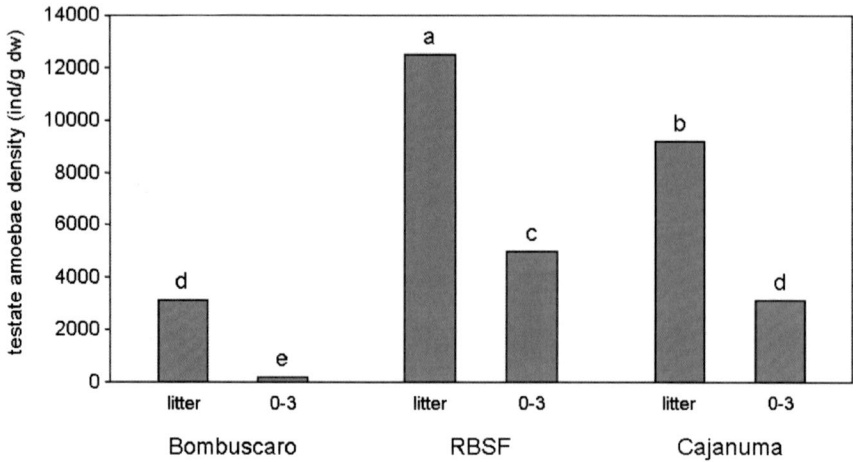

Figure 6. Testate amoebae densities at Bombuscaro (1000 m), RBSF (2000 m) and Cajanuma (3000 m) in the litter layer (litter) and the following deeper layer of a thickness of 3 cm (0-3). Bars sharing the same letter are not significantly different (Tukey's HSD test; P > 0.05; modified from Krashevska et al. 2007).

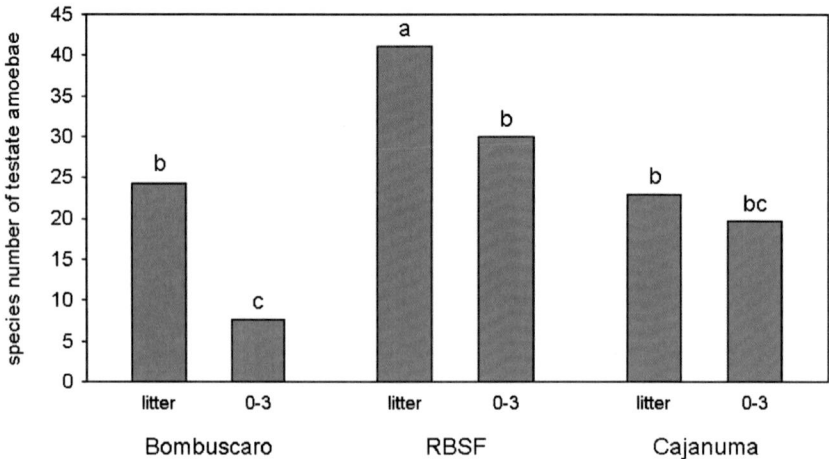

Figure 7. Number of species of testate amoebae at Bombuscaro (1000 m), RBSF (2000 m) and Cajanuma (3000 m) in the litter layer (litter) and the following deeper layer of a thickness of 3 cm (0-3). Bars sharing the same letter are not significantly different (Tukey's HSD test; P > 0.05).

dominant food substrate (Maraun et al. 2008; V. Krashevska, unpublished data).

A total of 135 species and intraspecific taxa of testate amoebae were found. Rarefaction plots suggest that only few more species are to be expected (Krashevska et al. 2007). Species number in the upper horizon was at a maximum at 2000 m and similar at 1000 and 3000 m (Fig. 7). The results suggest that species richness of testate amoebae does not decrease continuously with elevation; rather, it peaks at an intermediate elevation.

Morphological features of testate amoebae reflected semiaquatic habitat conditions. The great majority of testate amoebae species of the studied tropical mountain rainforests are geographically widespread, including temperate regions; however, nine of the species (i.e. 6.7%) are considered tropical, some of these species likely represent Gondwana relicts (Krashevska et al. 2007). There are contrasting views on the global distribution and diversity of protists (Finlay 2002, Foissner 2006). Generally, the observed pattern supports the view that (almost) 'everything is everywhere' since the great majority of the testate amoebae species recorded at the study sites also occurs in very different geographical regions, such as the temperate and boreal zone of the northern hemisphere (Krashevska et al. 2007).

Decomposition and microarthropod colonisation of litter. Litter decomposition in tropical mountain rainforests is slower than in lowland tropical rainforests (Heneghan et al. 1999) which may be due to lower temperatures at higher elevation and/or a decrease in litter quality with altitude. We studied the leaf litter decomposition of two abundant tree species (*Graffenrieda emarginata, Purdiaea nutans*) at the studied tropical mountain rainforest and evaluated the role of soil mesofauna for the decomposition of these two litter types (Illig et al. 2008). Both litter types decomposed slowly (average of 69% of initial dry weight remaining after 12 months; Fig. 8). The decomposition rates were similar to those of litter materials of forest ecosystems at higher latitude, e.g. oak-pine litter in a forest stand in Japan (Kaneko & Salamanca 1999), sweet chestnut and beech litter in forests in England (Anderson 1973) and evergreen oak litter in Northern Greece (Argyropoulou et al. 1993), but low compared to litter materials of tropical lowland forests. In these systems only about 30% of the initial dry weight remained after one year (Paoletti et al. 1991, Heneghan et al. 1999, Franklin et al. 2004).

One of the reasons for the slow decomposition may be low litter quality, i.e. low nutrient concentrations, in particular that of nitrogen (Enriques et al. 1993). Indeed, nitrogen concentrations of the leaf litter (L layer material) of *P. nutans* and *G. emarginata* were lower (C/N ratio of 73.6 and 41.9, respectively; J. Illig, unpubl. data) than those of L layer material of European beech (C/N ratio of 21; Maraun & Scheu 1996).

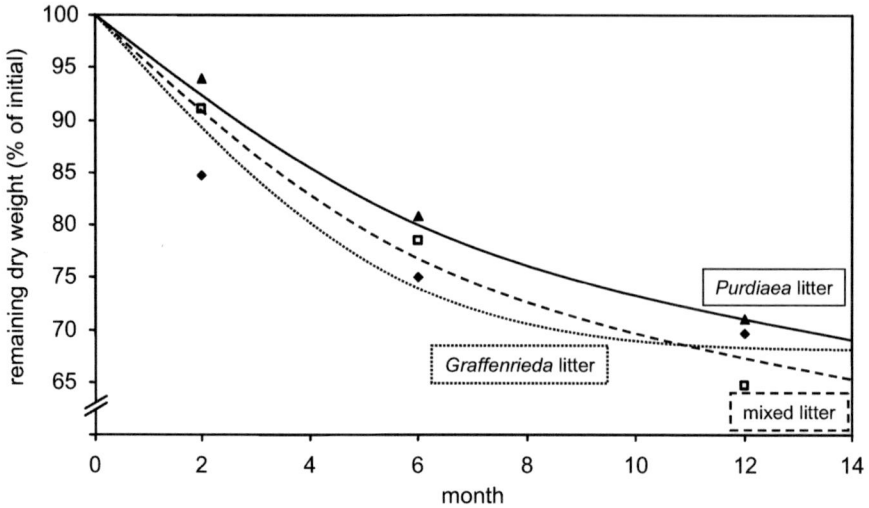

Figure 8. Decomposition (measured as remaining dry weight) of leaf litter of *Graffenrieda emarginata*, *Purdiaea nutans* and a mixture of both exposed in the field for 12 month (pooled data from exposure at 1850 and 2280 m; modified from Illig et al. 2008).

Each of the litter materials (*G. emarginata*, *P. nutans* and mixed litter) decomposed faster at 1850 m than at 2280 m (average of 60% and 76% of the initial litter mass remaining after 12 months, respectively). Lower decomposition rates at higher altitudes likely were due to lower temperatures, but increased nutrient limitation at higher altitude, in particular that of P (Maraun et al. 2008), may also have contributed to the decline in litter decomposition with altitude.

At the end of the experiment the mixture of *G. emarginata* and *P. nutans* litter had decomposed significantly faster than both single litter types, indicating that combining the two litter types accelerates decomposition processes (positive 'non-additive effect'). The reason for this non-additive litter decomposition may be the higher humidity in mixed litter (Wardle et al. 2003) or the mixing of litter of different quality which may increase the decomposition of the low quality litter, e.g. via nutrient translocation from high to low quality litter (Smith & Bradford 2003, Hättenschwiler et al. 2005, Quested et al. 2005).

The most abundant microarthropods in the litterbags were oribatid mites followed by Collembola, Gamasina, Uropodina, Prostigmata and Astigmata (Illig et al. 2008). Each of these taxa were more abundant at 1850 than at 2280 m. Oribatid mites in the litterbags constituted of 37 morphospecies and were dominated by *Scheloribates* sp., *Pergalumna sura* and *Truncozetes sturmi*. Species composition was similar in both litter types supporting previous findings that the structure of soil decomposer microarthropod communities is little affected by litter type (Walter

1985, Migge et al. 1998, Hansen 2000). As indicated by differences in litter decomposition between 48 μm and 1 mm mesh, soil microarthropods contributed little to litter decomposition. This is consistent with results of stable isotope based food web analyses suggesting that litter feeding soil meso- and macrofauna are scarce (Illig et al. 2005). However, higher density and diversity of secondary as compared to primary decomposers suggests that soil animals may affect decomposition processes at later stages of decay. Therefore, exposure of litter for 12 month, as done in the present study, may not have been long enough to evaluate the role of decomposer animals for litter decomposition in the studied tropical mountain rainforest. Long-term studies are needed to evaluate the role of soil fauna for litter decomposition in tropical mountain rainforests.

Conclusions

Contrary to our hypothesis, density and biomass of both soil meso- and microfauna at the studied tropical mountain rainforests are low as compared to temperate and boreal forest ecosystems. Similar to macrofauna, smaller soil invertebrates therefore presumably contribute little to litter decomposition processes, at least during early stages of litter decay. Consistent with the generally low density of decomposer animals, microbial biomass in litter and soil is also low (Maraun et al. 2008). Presumably, mircoorganisms and decomposer animals suffer from high energy demand due to high temperatures with the acquisition of energy being limited by low food quality (strong sclerotization of leaves, low nutrient content of litter materials, high abundance of lignified roots) in particular at high elevation. Consistent with low density, but contrary to animal groups above the ground, the species number of soil animals in the studied tropical mountain rainforests is similar to temperate forest ecosystems. This supports the assumption that the latitudinal gradient in diversity of belowground biota is less pronounced than above the ground. In contrast to temperate and boreal forest systems, where parthenogenetically reproducing soil animal species occur at high frequency, orbatid mites at the studied tropical mountain rainforests were dominated by sexually reproducing species in particular at higher altitude.

Overall, the results indicate that low density of soil animal taxa in tropical mountain rainforests is due to low resource quality and availability. Low frequency of parthenogenetic soil animal taxa in particular at higher elevation tropical rainforests supports this conclusion. Similar diversity of soil animal taxa in tropical as compared to temperate forest ecosystems indicates that trophic (and other) niches of decomposer species differ little between forests of different biomes. This supports the view that the diversity of plant species is of little importance for the diversity of belowground biota.

Acknowledgements. Financial support by the German Science Foundation (DFG) is gratefully acknowledged (FOR 402; MA 2461/2-1). Michael Heethoff helped in taking scanning electron microscope pictures of oribatid mites.

References

Anderson JM (1973) The breakdown and decomposition of sweet chestnut (*Castanea sativa* Mill) and beech (*Fagus sylvatica* L.) leaf litter in two deciduous woodland soils: I. Breakdown, leaching and decomposition. Oecologia 12: 251-274

Anichkin AE, Belyaeva NV, Dovgbrod IG, Shveenkova YB, Tiunov AV (2007) Soil microarthropods and macrofauna in monsoon tropical forests of cat Tien and Bi Dup-Nui Ba National Parks, Southern Vietnam. Biology Bulletin 34: 498-506

Argyropoulou MD, Asikidis MD, Iatrou GD, Stamou GP (1993) Colonization patterns of decomposing litter in a maquis ecosystem. European Journal of Soil Biology 29: 183-191

Brehm G, Fiedler K (2003) Faunal composition of geometrid moths changes with altitude in an Andean mountain rain forest. Journal of Biogeography 30: 431-440

Brehm G, Pitkin LM, Hilt N, Fiedler K (2005) Montane Andean rain forests are a global diversity hotspot of geometrid moths. Journal of Biogeography 32: 1621-1627

Domes K, Norton RA, Maraun M, Scheu S (2007) Re-evolution of sex in oribatid mites breaks Dollo's law. Proceedings of the National Academy of Sciences of the United States of America 104: 7139-7144

Enriques S, Duarte CM, Sand-Jensen K (1993) Pattern in decomposition rates among phytosynthetic organisms: the importance of detritus C:N:P content. Oecologia: 457-471

Finlay BJ (2002) Global dispersal of free-living microbial eukaryote species. Science 296: 1061-1063

Foissner W (2006) Biogeography and dispersal of micro-organisms: A review emphasizing protists. Acta Protozoologica 45: 111-136

Franklin E, Hayek T, Fagundes EP, Silva LL (2004) Oribatid mite (Acari: Oribatida) contribution to decomposition dynamic of leaf litter in primary forest, second growth, and polyculture in the central Amazon. Brazilian Journal of Biology 64: 59-72

Gauld ID, Gaston KJ, Janzen DH (1992) Plant alleclochemicals, tritrophic interactions and the anomalous diversity of tropical parasitoids – the nasty host hypothesis. Oikos 65: 353-357

Hamilton WD (1980) Sex versus non-sex versus parasite. Oikos 35: 282-290

Hansen RA (1999) Red oak litter promotes a microarthropod functional group that accelerates its decomposition. Plant and Soil 209: 37-45

Hansen RA (2000) Effects of habitat complexity and composition on a diverse litter microarthropod assemblage. Ecology 81: 1120-1132

Hättenschwiler S, Tiunov AV, Scheu S (2005) Biodiversity and litter decomposition in terrestrial ecosystems. Annual Review of Ecology, Evolution and Systematics 36: 191-218

Heneghan L, Coleman DC, Zou X, Crossley DA, Haines BL (1999) Soil microarthropod contributions to decomposition dynamics: tropical-temperate comparisons of a single substrate. Ecology 80: 1873-1882

Homeier J, Dalitz H, Breckle SW (2002) Waldstruktur und Baumartendiversität im montanen Regenwald der Estation Cientifica San Francisco in Südecuador. Berichte der Reinhard-Tüxen-Gesellschaft 14: 109-118

Illig J, Langel R, Norton RA, Scheu S, Maraun M (2005) Where are the decomposers? Uncovering the soil food web of a tropical mountain rain forest in southern Ecuador using stable isotopes (^{15}N). Journal of Tropical Ecology 21: 589-593

Illig J, Sandmann D, Schatz H, Scheu S, Maraun M (2007) Oribatida (Mites) - Checklist Reserva Biologica San Francisco (Prov. Zamora-Chinchipe, S. Ecuador). Ecotropical Monographs 4: 221-230

Illig J, Schatz H, Scheu S, Maraun M (2008) Decomposition rates and micro-arthropod colonization of litterbags with different litter types (*Graffenrieda emarginata, Purdiaea nutans*) in a tropical mountain rain forest in southern Ecuador. Journal of Tropical Ecology 24, in press

Kaneko N, Salamanca E (1999) Mixed leaf litter effects on decomposition rates and soil microarthropod communities in an oak-pine stand in Japan. Ecological Research 14: 131-138

Kempson D, Lloyd M, Ghelardi R (1963) A new extractor for woodland litter. Pedobiologia 3: 1-21

Kessler M, Kessler PJA, Gradstein SR, Bach K, Schmull M, Pitopang R (2005) Tree diversity in primary forest and different land use systems in Central Sulawesi, Indonesia. Biodiversity and Conservation 14: 547-560

Krashevska V, Bonkowski M, Maraun M, Scheu S (2007) Testate amoebae (Protista) of an elevational gradient in the tropical mountain rain forest of Ecuador. Pedobiologia 51: 319-331

Lucky A, Erwin TL, Witman JD (2002) Temporal and spatial diversity and distribution od arboreal Carabidae (Coleoptera) in a western Amazonian rain forest. Biotropica 34: 376-386

Maraun M, Scheu S (1996) Seasonal changes in micorbial biomass and activity in leaf litter layers of beech (*Fagus sylvatica*) forests on a basalt-limestone gradient. Pedobiologia 40: 21-31

Maraun M, Scheu S (2000) The structure of oribatid mite communities (Acari, Oribatida): patterns, mechanisms and implications for future research. Ecography 23: 374-383

Maraun M, Heethoff M, Scheu S, Weigmann G, Norton RA, Thomas RH (2003a) Radiation in sexual and parthenogenetic oribatid mites (Oribatida, Acari) as indicated by genetic divergence of closely related species. Experimental and Applied Acarology 29: 265-277

Maraun M, Salamon JA, Schneider K, Schaefer M, Scheu S (2003b) Oribatid mite and collembolan diversity, density and community structure in a moder beech forest (*Fagus sylvatica*): effects of mechanical disturbances. Soil Biology and Biochemistry 35: 1387-1394

Maraun M, Schatz H, Scheu S (2007) Awesome or ordinary? Global diversity patterns of oribatid mites. Ecography 30: 209-216

Maraun M, Illig J, Sandmann D, Krashevska V, Norton RA, Scheu S (2008) Soil Fauna. In: Beck E, Bendix J, Kottke I, Makeschin F, Mosandl R (eds) Gradients in a Tropical Mountain Ecosystem of Ecuador. Ecological Studies 198. Springer, Berlin, Heidelberg, New York, pp 181-192

Migge S, Maraun M, Scheu S, Schaefer M (1998) The oribatid mite community (Acarina) on pure and mixed stands of beech (*Fagus sylvatica*) and spruce (*Picea abies*) at different age. Applied Soil Ecology 9: 119-126

Myers M, Mittermeier RA, Mittermeier CG, da Fonseca GAB, Kent J (2000) Biodiversity hotspots for conservation priorities. Nature 403: 853-858

Niedbala W, Illig J (2007a) Ptyctimous mites (Acari, Oribatida) from the Ecuador rainforest. Journal of Natural History 41: 771-777

Niedbala W, Illig J (2007b) New species of ptyctimous mites (Acari, Oribatida) from Ecuador. Tropical Zoology 20: 135-150

Oosterhoorn M, Kappelle M (2000) Vegetation structure and composition along an interior-edge-exterior gradient in a Coast Rica mountain cloud forest. Forest Ecology and Management 126: 291-307

Opravilova V, Hajek M (2006) The variation of testacean assemblages (Rhizopoda) along the complete base-richness gradient in fens: A case study from the Western Carpathians. Acta Protozoologica 45: 191-204

Otto SP, Nuismer SL (2004) Species interactions and the evolution of sex. Science 304: 1018-1020

Paoletti MG, Taylor RAJ, Stinner BR, Stinner DH, Benzing DH (1991) Diversity of soil fauna in the canopy and forest floor of Venezuelan cloud forest. Journal of Tropical Ecology 7: 373-383

Plowman KP (1981) Distribution of Cryptostigmata and Mesostigmata (Acari) within the litter and soil layers of two subtropical forests. Australian Journal of Ecology 6: 365-374

Quested HM, Callaghan TV, Cornelissen JHC, Press MC (2005) The impact of hemiparasitic plant litter on decomposition: Direct, seasonal and litter mixing effects. Journal of Ecology 93: 87-98

Röderstein M, Hertel D, Leuschner C (2006) Above- and below-ground litter production in three tropical montane forests in southern Ecuador. Journal of Tropical Ecology 21: 483-492

Schaefer M (1991) Animals in European temperate deciduous forests. In: Röhrig E, Ulrich B (eds) Temperate deciduous forests (Ecosystems of the World). Elsevier, Amsterdam, pp 503-525

Schaefer M (1999) The diversity of the fauna of two beech forests: some thoughts about possible mechanisms causing the observed patterns. In: Kratochwil A (ed) Biodiversity in ecosystems: Principles and case studies of different complexity levels. Kluwer Academic Publishers, Dordrecht, pp 45-64

Schatz H (2008) Biogeography of oribatid mites (Acari, Oribatida) from the Cordillera de Talamanca, Costa Rica and Panama. In: Morales-Malacara JB, Behan-Pelletier V, Ueckermann E, Pérez TM, Estrada E, Gispert C, Badii M (eds) Acarology XI: Proceedings of the International Congress, in press

Scheu S (2005) Linkages between tree diversity, soil fauna and ecosystem processes. In: Scherer-Lorenzen M, Körner C, Schulze ED (eds) Forest diversity and function: Temporate and boreal systems. Ecological Studies 176, Springer, Berlin, pp 211-233

Scheu S, Drossel B (2007) Sexual reproduction prevails in a world of structured resources in short supply. Proceedings of the Royal Society B – Biological Sciences 274: 1225-1231

Schneider K, Migge S, Norton RA, Scheu S, Langel R, Reineking A, Maraun M (2004) Trophic niche differentiation in oribatid mites (Oribatida, Acari): evidence from stable isotope ratios ($^{15}N/^{14}N$). Soil Biology and Biochemistry 36: 1769-1774

Schönborn W (1988). Biometric studies on species, races, ecophenotypes and individual variations of soil-inhabiting Testacea (Protozoa, Rhizopoda), including *Trigonopyxis minuta* n sp and *Corythion asperulum* n sp. Archiv für Protistenkunde 136: 345-363.

Smith VC, Bradford MA (2003) Do non-additive effects on decomposition in litter-mix experiments result from differences in resource quality between litters? Oikos 102: 235-242

Soethe N, Lehmann J, Engels C (2006) The vertical pattern of rooting and nutrient uptake at different altitudes of a south Ecuadorian mountain forest. Plant and Soil 286: 287-299

Valkanov A (1962) *Euglyphella delicatula* n g, n sp. (Rhizopoda, Testacea) und ihre Kopulation. Comptes Rendus, Academie Bulgare des Sciences 15: 207-209.

Walter DE (1985) The effect of litter type and elevation on colonization of mixed coniferous litterbags by oribatid mites. Pedobiologia 28: 383-387

Wardle DA, Nilsson MC, Zackrisson O, Gallet C (2003) Determinants of litter mixing effects in a Swedish boreal forest. Soil Biology and Biochemistry 35: 827-835

Whitmore TC (1998) An introduction to tropical rain forests. Oxford University Press, Oxford.

Wilcke W, Yasin S, Abramowski U, Valarezo O, Zech W (2002) Nutrient storage and turnover in organic layers under tropical mountain rain forest in Ecuador. European Journal of Soil Science 53: 15-27

Biodiversity and Ecology Series (2008) 2: 97-107
The Tropical Mountain Forest – Patterns and Processes in a Biodiversity Hotspot
edited by S.R. Gradstein, J. Homeier and D. Gansert
Göttingen Centre for Biodiversity and Ecology

The influence of topography on forest structure and regeneration dynamics in an Ecuadorian montane forest

Jürgen Homeier

Albrecht von Haller Institute of Plant Sciences, Department of Plant Ecology, University of Göttingen, Untere Karspüle 2, 37073 Göttingen, Germany, jhomeie@gwdg.de

Abstract. The forests of the Reserva Biológica San Francisco (RBSF) in the southern Ecuadorian Andes constitute a hotspot of biodiversity. One of the causes of the high taxonomic diversity is the great topographic heterogeneity of the area. Structure, tree species composition and tree regeneration of ravine and ridge forests in the RBSF were studied to better understand the mechanisms underlying topographical changes in vegetation. Differences in species composition were reflected in the seedling communities of the two forest types. Ravine forests were generally more productive (as expressed by basal area growth) and showed higher tree species richness. Soil nutrient concentration seems to be a major factor driving forest structure and tree species composition in the study area. The topographical gradient with its array of edaphic conditions thus contributes to the variety of (micro-)habitats and increases vegetation heterogeneity and plant species diversity of the studied montane forest.

Introduction

Vegetation structure and species composition of tropical forests characteristically undergo changes along altitudinal gradients. An overall decrease in tree species richness, forest stature and above ground primary productivity with increasing elevation has often been documented (e.g. Aiba & Kitayama 1999, Grubb 1977, Homeier et al. 2008, Leuschner et al. 2007, Lieberman et al. 1996), and is shown for Ecuadorian forests in Fig. 1. The graph shows a clear decline in tree species diversity with increasing elevation, from lowland to upper montane forests.

In addition to elevation, topography is an important factor affecting vegetation structure and species diversity, by providing (micro-)habitat heterogeneity (e.g. Clark et al. 1999, Harms et al. 2001, Lieberman et al. 1985, Phillips et al. 2003, Webb et al. 1999). Some authors have suggested that changes in species composition and forest structure along topographical gradients are similar to those along elevational gradients (e.g. Tanner 1977, Webb et al. 1999, Takyu et al. 2002).

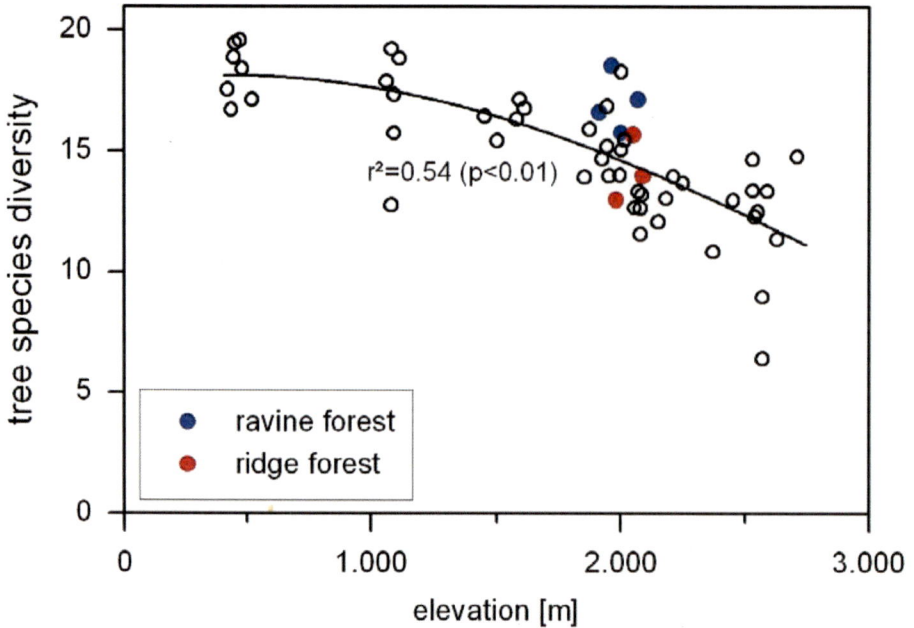

Figure 1. Tree species diversity (shown as rarefied species for n=21 trees within an area of 400 m²) of 55 Ecuadorian primary forest stands. Ridge and ravine forests of the Reserva Biológica San Francisco are marked.

Topography is a complex factor and relates to hydrology, nutrient dispersion, soil structure and wind exposure, which are difficult to disentangle. As a consequence, the mechanisms underlying elevational and topographical changes of plant diversity and forest structure are still poorly understood, and studies documenting horizontal gradients and their effects on vegetation in tropical mountains are still scarce (e.g. Luizao et al. 2004, Takyu et al. 2002).

The montane forests of the Río San Francisco valley in southern Ecuador

The forests of the Reserva Biológica San Francisco (RBSF) in the southern Ecuadorian Andes have recently been subject of detailed ecological study (e.g. Beck et al. 2008, Liede & Breckle 2008). The area is a hotspot of biodiversity and has a very high diversity of vascular plants (1208 species of spermatophytes, 257 of ferns; Homeier & Werner 2008, Lehnert et al. 2008, see also chapter 13 of this volume). Diversity of bryophytes (> 500 species per 1000 ha; Parolly et al. 2004, Gradstein et al. 2008) and geometrid moths (> 1000 species; Brehm et al. 2005) is

Figure 2. The Reserva Biológica San Francisco, southern Ecuador, is characterized by rugged topography. Landslides are an important factor in vegetation dynamics.

the highest ever recorded for an area of comparable size. One of the causes of the high taxonomic diversity is the great topographic heterogeneity of the region (Homeier et al. 2008). Most of the area of the RBSF is situated on the steeply sloping (20-50°), north-facing flank of the Río San Francisco valley consisting of a complex system of ridges, steep slopes, valleys and ravines. The topography is generally very steep and rugged, and soils are highly heterogeneous, being richer in nutrients in ravines than on ridges and showing an overall trend towards more unfavourable conditions for plant growth with elevation (Homeier 2004, Soethe et al. 2008, Wilcke et al. in press) (Fig. 2).

More than 280 tree species are currently known from the RBSF. Four main types of primary forest vegetation have been distinguished (Homeier et al. 2008), two of which are briefly described here. The tallest and most speciose forest type (type I) is found at 1850–2200 m on the lower slopes of the Río San Francisco valley and along ravines, with a canopy attaining 25–30 m and some emergent trees reaching up to 35 m. Megaphyllous shrubs (Piperaceae) and herbs (Zingiberaceae, Heliconiaceae) are common in the understorey. Important tree families are Melastomataceae, Rubiaceae, Lauracae and Euphorbiaceae; characteristic taxa include *Tabebuia chrysantha* (Bignoniaceae), *Hyeronima asperifolia* (Euphorbiaceae),

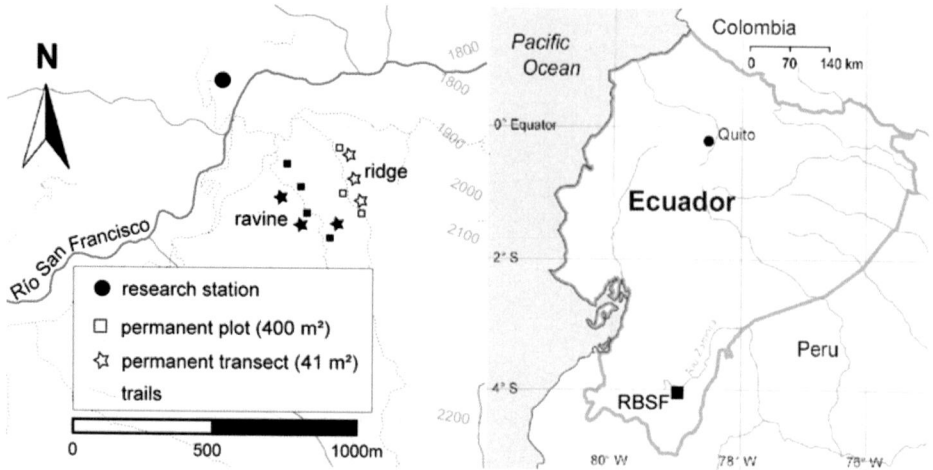

Figure 3. The study area in the Reserva Biológica San Francisco (RBSF) with the location of the primary forest plots and the forest regeneration transects. Three plots and three transects are situated on a gentle ridge (white circles/white stars), four plots and three transects within a ravine (black circles/black stars).

Meriania sp. and *Miconia* spp. (Melastomataceae), *Inga* spp. (Mimosaceae), *Piper* spp. (Piperaceae) and *Micropholis guyanensis* (Sapotaceae). Along ridges and upper slopes at the same elevation forest stature and tree species composition differ significantly, with only few trees reaching more than 15 m in height (forest type II). Characteristic and frequent tree species include *Alzatea verticillata* (Alzateaceae), *Hyeronima moritziana* (Euphorbiaceae), *Graffenrieda emarginata* (Melastomataceae), *Podocarpus oleifolius* (Podocarpaceae), *Matayba inelegans* (Sapindaceae) and various species of Lauraceae.

Influence of topography on forest structure

The influence of topography on forest structure was studied using permanent plots of 400 m² in the two forest types (Homeier 2004; Fig. 3). Ravines (forest type I) and ridges (forest type II) differed notably with respect to structural parameters (Fig. 4). Canopy openness was significantly higher in ridge forest where trees were shorter and stem density higher than in ravine forest. A decrease in stem slenderness, as typical for altitudinal gradients, was not observed from ravine to ridge forest. Tree basal area was higher in ravine forest, as were annual basal area growth, stem density and species density (Fig. 1). Trees in ridge forest were generally slow-growing while those in ravine forest showed a wide range of growth rates (Homeier 2004).

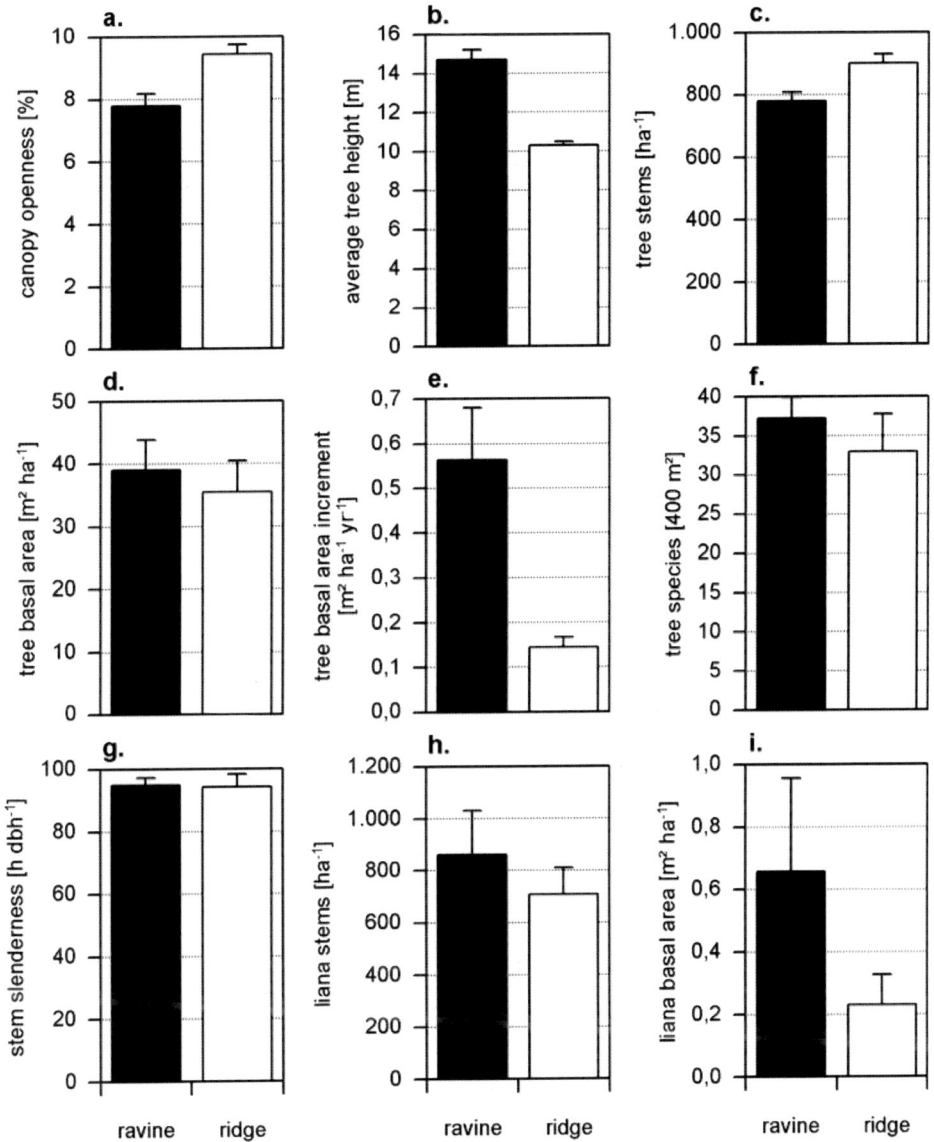

Figure 4. Structural parameters of ravine and ridge forest in the Reserva Biológica San Francisco. Shown are means and standard errors, calculated for all trees with dbh $\geq$ 10 cm (dbh $\geq$ 5 cm in g.) and climbers with dbh $\geq$ 1 cm from the permanent forest plots (Fig. 3). Data of parameters b and c are taken from Homeier et al. (in prep.). Differences in parameters a, b, c, f, and i are significant (U-test, $p < 0.05$).

Density of climbers was only slightly higher in ravines than on ridges but climber basal area was markedly higher in ravines, where large diameters were much more common. The structural differences between the forest types of the RBSF correlated with differences in soil nutrient concentration, which was highest in ravine forest (Homeier 2004, Wilcke et al., in press). Oesker et al. (in press) compared nutrient concentrations of throughfall collected in ravine forest and ridge forest and found that nutrient contents in ravine forest are generally higher, thus reflecting the more favourable soil conditions of the ravines. Because there are no other obvious differences in site conditions, the soil nutrient concentration appears to be the major factor governing tree species composition and growth processes in the two forest types.

Tree regeneration in ravine forest and ridge forest

Since seedlings usually react more sensitively to environmental parameters than mature plants, monitoring of primary forest regeneration is a key approach to the study of the mechanisms underlying topographical changes in vegetation. I studied tree regeneration along six cross-shaped transects (of 41 m²), three in each of the two above-mentioned forest types (Homeier & Breckle 2008; Fig. 3). Transects were situated at the forest-gap transition with about half of the transect area below closed canopy and the rest in a recent gap (= natural tree-fall). By repeated inventorying of tree seedlings and saplings taller than 5 cm during five years (2003-2007), spatial and topographical regeneration dynamics of tree species were analysed.

Seedling species composition of the two forest types was quite different (Table 1). Some families such as Arecaceae and Melastomataceae occurred principally in ridge forest whereas others like Piperaceae and Rubiaceae were more common in the ravines. Mimosaceae were found almost exclusively in ravine forest. Regeneration of ravine forests showed distinct succession phases similar to those reported from tropical lowland forests (e.g. Whitmore 1989, 1996). After gap formation, fast-growing pioneer tree species (e.g. *Piptocoma discolor*, *Cecropia andina*, *Heliocarpus americanus*) together with seedlings and saplings surviving gap formation formed a closed canopy. Subsequently, the short-lived pioneers are replaced by late-successional species germinating below the closed canopy. Together with long-lived pioneers such as *Tabebuia chrysantha* and *Cedrela* sp., the late-successional trees build the canopy of the mature ravine forest.

In contrast, species composition of early and late successional phases of ridge forest was rather uniform, and common ravine forest pioneers were completely lacking. Their absence may be explained by dispersal limitation in the case of animal-dispersed species (e.g. *Cecropia andina*), but the immediate vicinity of the two forest types makes dispersal constraints unlikely. Rather it is likely to be the result

Table 1. The ten most important families of tree species seedlings regenerating in ravine and ridge forest in the Reserva Biológica San Francisco, southern Ecuador. Figures are relative abundance values for the two forest types and overall means.

family	ravine [%]	ridge [%]	total [%]
Melastomataceae	14.2	27.9	23.6
Rubiaceae	24.8	17.0	19.4
Arecaceae	4.8	13.4	10.7
Lauraceae	7.8	10.3	9.5
Mimosaceae	10.1	0	3.2
Piperaceae	6.9	1.1	2.9
Euphorbiaceae	0.8	2.6	2.0
Araliaceae	1.9	1.8	1.9
Cyatheaceae	0.8	2.3	1.8
Asteraceae	2.4	0.5	1.1
others	25.6	23.2	23.9

of unfavourable germination conditions, particularly with regard to wind-dispersed species (*Cedrela* sp., *Heliocarpus americanus*, *Piptocoma discolor*, *Tabebuia chrysantha*).

The number of seedling individuals in the ravine transects was consistently lower than in the ridge transects. Average seedling density was more than twofold higher in ridge forest (Fig. 5). In contrast, ravine plants on average were significantly taller than ridge forest plants in all inventories. Relative height increments showed highest variation within the smaller size classes of woody regeneration at both sites. On average, plants at both sites grew ca. 5 cm in height in one year (Homeier & Breckle 2008). Average growth of regenerating plants showed no differences between ravine and ridge, as different from tree basal area. The higher stem density and smaller tree size in ridge forest as compared with ravine forest is maintained until the mature forest phase. These structural differences seem to be largely caused by species composition.

Because trees in ravine forest are larger in stature, maximum stem density in ravine forest is lower than in ridge forest. On the other hand, the more favourable light conditions in ridge forest due to the lower and more open canopy may promote higher seedling densities in the latter forest type. Also, mortality and recruitment of seedlings and saplings were significantly higher on ridges than in ravines (Fig. 6), resulting in higher dynamics of seedling populations in ridge forest. In general, mortality at both sites was highest among the smallest plants and decreased with seedling size (Homeier & Breckle 2008). An extreme increase in mortality observed in the last inventory (2007) was probably due to an unusual drought event in 2006.

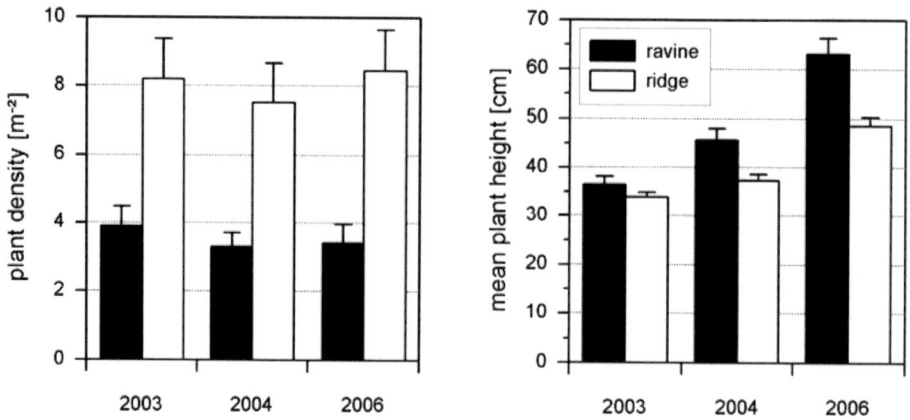

Figure 5. Mean plant density (*left*) and mean height (*right*) in the forest regeneration transects in the three inventories. Given are means and standard errors.

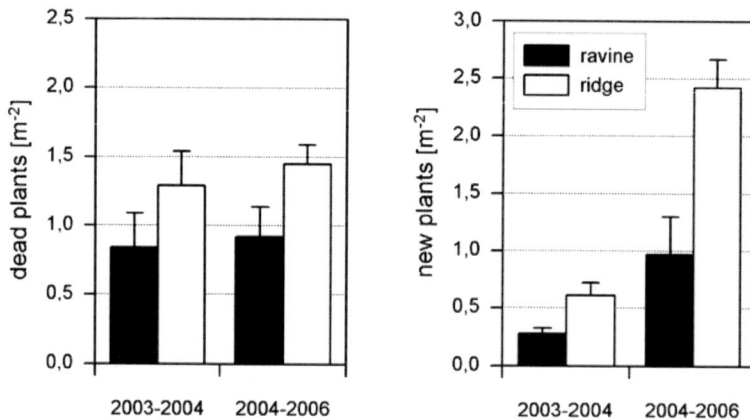

Figure 6. Mean plant mortality (*left*) and mean recruitment (*right*) in the studied forest regeneration transects for two consecutive periods of one year and two years, respectively. Given are means and standard errors.

Conclusions

The present study shows that patterns of woody plant regeneration in tropical montane forest can be quite different among ridge and ravine forests. Differences in tree composition and population dynamics in the two neighbouring forest types are maintained during the succession from young towards mature forest. Ravine forests are generally more productive (as expressed by basal area growth) and show higher tree species richness. Based on the current knowledge of the RBSF, soil nutrient concentration seems to be one major underlying factor driving forest structure and tree species composition. The different structure is principally caused by differences in tree species composition, which most likely seems to be a result of recruitment limitation in ridge forests. The topographical gradient with its array of edaphic conditions thus contributes to the variety of (micro-) habitats and increases vegetation heterogeneity and plant species diversity of the studied montane forest.

Acknowledgements. The author thanks Melanie Wachter, Elke Brandes, Rebecca Scetaric, Friedrich Angermüller, Anna-Maria Weißer, Pascal Hecht, and, especially, Lars Nauheimer for helping with regeneration monitoring, the Ministerio del Ambiente for research permits, and the German Research Foundation (DFG - FOR 402, FOR 816) for financial support.

References

Aiba S, Kitayama K (1999) Structure, composition and species diversity in an altitude-substrate matrix of rain forest tree communities on Mount Kinabalu, Borneo. Plant Ecology 140: 139-157

Beck E, Bendix J, Kottke I, Makeschin F, Mosandl R (2008) Gradients in a Tropical Mountain Ecosystem of Ecuador. Ecological Studies 198, Berlin, Heidelberg, New York

Brehm G, Pitkin LM, Hilt N, Fiedler K (2005) Montane Andean rain forests are a global diversity hotspot of geometrid moths. Journal of Biogeography 32: 1621-1627

Clark DB, Palmer MW, Clark DA (1999) Edaphic factors and the landscape-scale distributions of tropical rain forest trees. Ecology 80: 2662-2675

Gradstein SR, Bock C, Mandl N, Nöske N (2008) Bryophyta: Liverworts and hornworts. In: Liede-Schuhmann S, Breckle SW (eds) Provisional Checklists of Flora and Fauna of the San Francisco valley and its surroundings, Southern Ecuador. Ecotropical Monographs 4: 69-87

Grubb PJ (1977) Control of forest growth and distribution on wet tropical mountains, with special reference to mineral nutrition. Annual Review of Ecology and Systematics 8: 83-107

Harms K, Condit R, Hubbell SP, Foster RB (2001) Habitat association of trees and shrubs in a 50-ha neotropical forest plot. Journal of Ecology 89: 947-959

Homeier J (2004) Baumdiversität, Waldstruktur und Wachstumsdynamik zweier tropischer Bergregenwälder in Ecuador und Costa Rica. Dissertationes Botanicae 391: 1-207.

Homeier J, Breckle S-W (2008) Gap-dynamics in a tropical lower montane forest in South Ecuador. In: Beck E, Bendix J, Kottke I, Makeschin F, Mosandl R (eds) Gradients in a Tropical Mountain Ecosystem of Ecuador. Ecological Studies 198. Springer, Berlin, Heidelberg, New York, pp 311-317

Homeier J, Werner FA (2008) Spermatophyta. In: Liede-Schuhmann S, Breckle SW (eds) Provisional Checklists of Flora and Fauna of the San Francisco valley and its surroundings, Southern Ecuador. Ecotropical Monographs 4: 15-58

Homeier J, Werner FA, Gradstein SR, Breckle S-W, Richter M (2008) Potential vegetation and floristic composition of Andean forests in South Ecuador, with a focus on the RBSF. In: Beck E, Bendix J, Kottke I, Makeschin F, Mosandl R (eds) Gradients in a Tropical Mountain Ecosystem of Ecuador. Ecological Studies 198. Springer, Berlin, Heidelberg, New York, pp 87-100

Lehnert M, Kessler M, Salazar LI, Navarrete H, Werner FA, Gradstein SR (2008) Spermatophyta. In: Liede-Schuhmann S, Breckle SW (eds) Provisional Checklists of Flora and Fauna of the San Francisco valley and its surroundings, Southern Ecuador. Ecological Monographs 4: 59-68

Leuschner C, Moser G, Bertsch C, Röderstein M, Hertel D (2007) Large altitudinal increase in tree shroot/shoot ratio in tropical mountain forests of Ecuador. Basic and Applied Ecology 8: 219-230

Lieberman M, Lieberman, D, Hartshorn, GS, Peralta R (1985) Small-scale altitudinal variation in lowland wet tropical forest vegetation. Journal of Ecology 73: 505-516

Lieberman D, Lieberman M, Peralta R, Hartshorn GS (1996) Tropical forest structure and composition on a large-scale altitudinal gradient in Costa Rica. Journal of Ecology 84: 137-152

Liede-Schuhmann S, Breckle SW (2008) Provisional Checklists of Flora and Fauna of the San Francisco valley and its surroundings (Estación Científica San Francisco), Southern Ecuador. Ecotropica Monographs 4

Luizao RCC, Luizao FJ, Paiva RQ, Monteiro TF, Sousa LS, Kruijt B (2004) Variation of carbon and nitrogen cycling processes along a topographic gradient in a central Amazonian forest. Global Change Biology 10: 592-600

Oesker M, Homeier J, Dalitz H (in press) Spatial heterogeneity of canopy throughfall quantity and quality in a tropical montane forest in South Ecuador. In: Bruijnzeel S, Juvik J (eds) Mountains in the mist: Science for Conserving and Managing tropical montane cloud forests. Hawaii University Press

Parolly G, Kürschner H, Schäfer-Verwimp A, Gradstein SR (2004) Cryptogams of the Reserva Biológica San Francisco (Province Zamora-Chinchipe, Southern Ecuador) III. Bryophytes - Additions and new species. Cryptogamie, Bryologie 25: 271-289

Phillips OL, Nunez Vargas P, Lorenzo Monteagudo A, Pena Cruz A, Chuspe Zans M-E, Galiano Sanchez W, Yli-Halla M, Rose S (2003) Habitat association among Amazonian tree species: a landscape-scale approach. Journal of Ecology 91: 757–775

Soethe N, Wilcke W, Homeier J, Lehmann J, Engels C (2008) Plant growth along the altitudinal gradient – role of plant nutritional status, fine root activity and soil properties. In: Beck E, Bendix J, Kottke I, Makeschin F, Mosandl R (eds) Gradients in a Tropical Mountain Ecosystem of Ecuador. Ecological Studies 198. Springer, Berlin, Heidelberg, New York, pp 259-266

Takyu M, Aiba S-I, Kitayama K (2002) Effects of topography on tropical lower montane forests under different geological conditions on Mount Kinabalu, Borneo. Plant Ecology 159: 35-49

Tanner EVJ (1977) Four montane rain forests of Jamaica: A quantitative characterization of the floristics, the soils and the foliar mineral levels, and a discussion of the interrelations. Journal of Ecology 65: 883-918

Webb EL, Stanfield, BJ, Jensen ML (1999) Effects of topography on rainforest tree community structure and diversity in American Samoa, and implications for frugivore and nectarivore populations. Journal of Biogeography 26: 887-897

Whitmore TC (1989) Canopy gaps and the two major groups of forest trees. Ecology 70: 536-538

Whitmore TC (1996) A review of some aspects of tropical rain forest seedling ecology with suggestions for further enquiry. In: Swaine MD (ed) The ecology of tropical forest tree seedlings. Man and the biosphere series 17. UNESCO, Paris, pp 3-39

Wilcke W, Oelmann Y, Schmitt A, Valarezo C, Zech W, Homeier J (in press) Soil properties and tree growth along an altitudinal transect in Ecuadorian tropical montane forest. Journal of Plant Nutrition and Soil Science

Biodiversity and Ecology Series (2008) 2: 109-128
The Tropical Mountain Forest – Patterns and Processes in a Biodiversity Hotspot
edited by S.R. Gradstein, J. Homeier and D. Gansert
Göttingen Centre for Biodiversity and Ecology

Carbon allocation and productivity in tropical mountain forests

Christoph Leuschner and Gerald Moser

Albrecht von Haller Institute of Plant Sciences, Department of Plant Ecology, University of Göttingen, Untere Karspüle 2, 37073 Göttingen, Germany, cleusch@uni-goettingen.de

Abstract. Tropical rainforests decrease in tree height and aboveground biomass (AGB) with increasing elevation. The causes of this phenomenon remain insufficiently understood. We report about a transect study in southern Ecuador which consisted of five forest stands along an altitudinal transect from 1050 to 3060 m. We investigated stand structure, aboveground and belowground biomass, leaf mass and area, and leaf mass production. Average tree height and AGB were reduced to less than 50% between 1050 and 3060 m, LAI decreased from 6.0 to 2.2. The leaf area reduction must have resulted in a lowered canopy carbon gain and thus may partly explain the reduced tree growth at high elevation. On the other hand, fine and coarse root biomass both significantly increased with elevation across this transect. The ratio of root biomass (fine and coarse) to AGB increased greatly from 1050 to 3060 m. Under the assumption that fine root biomass does reflect productivity, our data indicate a marked belowground shift in carbon allocation with increasing elevation. Possible explanations for this allocation shift are discussed including reduced N supply due to low temperatures, water logging and adverse soil chemical conditions.

Introduction

Perhaps the most obvious change that occurs with elevation on tropical mountains is a gradual decrease in tree height (Whitmore 1998). In equatorial regions, tropical lowland forest is replaced by Lower Montane Forest (LMF) typically at about 1000–1200 m a.s.l., giving way to Upper Montane Forest (UMF) at about 1800–2200 m. LMF trees reach a maximum height of 25–35 m (at the most 40 m), whereas UMF trees do not grow taller than 18–22 m and decrease in height toward the tree line where less than 10 m are reached. Elfin forests with trees 3–7 m tall are common immediately below the tree line in many high equatorial mountains. Average leaf size and specific leaf area decrease with elevation as well (Ashton 2003).

For several decades, there has been an intensive debate among plant ecologists on the causes of this altitudinal reduction in tree size in tropical mountains (see also chapter 10 of this volume). A number of explanations for reduced growth in tropical high-elevation forests has been suggested including reduced plant surface temperatures due to dense cloud cover (Grubb 1977), periodic drought stress on shallow mountain soils (van Steenis 1972), limited oxygen supply in water-logged soils (Hetsch & Hoheisel 1976), low nutrient supply (Tanner et al. 1998), limited nutrient uptake due to diminished transpiration, high concentrations of phenolic compounds and/or free aluminium in soil organic matter (Bruijnzeel & Proctor 1995, Bruijnzeel & Veneklaas 1998, Hafkenscheid 2000), strong winds (Sugden 1986), or elevated UV-B radiation and damage of the photosynthetic apparatus at high elevation (Flenley 1995). An alternative hypothesis is that the altitudinal decrease in tree height in tropical mountains is primarily the consequence of a shift in aboveground/belowground carbon allocation patterns, whereas sink or source limitation of carbon turnover are of only secondary importance. There are recent observations showing a much larger fine root biomass (FRB) in tropical mountain forests than in lowland forests (Cairns et al. 1997, Hertel et al. 2003, Kitayama & Aiba 2002, Vogt et al. 1996). However, sound conclusions on altitudinal trends in tree root biomass of tropical mountain forests cannot be drawn because the data is very limited. Moreover, the underlying causes of putative allocation shifts remain unclear.

Material and Methods

Study sites. Five mountain forest stands in the South Ecuadorian province of Zamora-Chinchipe on the eastern slopes of the Andes were selected for comparative study. The study area is one of the remaining regions in the northern Andes with only low human impact on forest vegetation in a broad altitudinal zone. Stand selection was based on the following attributes: (a) no visible disturbance by humans or landslides, (b) more or less closed canopies without major treefall gaps, (c) structural homogeneity in an area of at least 100 m x 100 m, and (d) position on moderately steep slopes (10–30°) in a more or less smooth terrain, thus avoiding hollows and ridges. The study plots were selected in appropriate forest sections that fulfilled the above criteria.

The five stands are located on the northern slopes of the Podocarpus National Park between Loja and Zamora (S 03°58'–04°06', W 78°58'–79°10') at a maximum distance to each other of 30 km. With elevations of 1050, 1540, 1890, 2380 and 3060 m a.s.l., an altitudinal gradient of 2000 m was covered. Stands 1 and 2 (1050 and 1540 m) are 10 km south of Zamora upslope of the village of Bombuscaro, stand 3 (1890 m) is in close vicinity of Estación Científica San Francisco (ECSF) south of the road from Loja to Zamora. Stand 4 (2380 m) is located 400 m upslope of ECSF, stand 5 (3060 m) in the Cajanuma area at the northwestern gate of the

National Park about 20 km south of Loja. The stands are situated on slopes facing east to northwest.

While the lowermost stand (1050 m) is situated in the transition zone between tropical lowland and LMF, the stands 2, 3 and 4 are in the lower to upper montane belts (bosque siempreverde montano, Balslev & Øllgaard 2002; bosque de neblina montano, Valencia et al. 1999). Stand 5 (3060 m) is situated close to the tree line which is found at 3200–3400m in the Loja/Zamora region. This stand is a typical 'elfin forest' with stunted tree growth. Patches of alpine páramo are found about 200 m upslope of this site. The structure and floristics of the forests at the five sites are considerably different. However, the same plant life form (i.e. evergreen broad-leaved trees) occurs throughout. All forests are species-rich with about 100 tree species per 0.6 ha. Lauraceae, Melastomataceae and Rubiaceae are the most species-rich and most frequent tree families in the stands (Homeier 2004).

The soils have formed on acidic phyllite, sandstone or shale bedrocks with Alumic Acrisols being present at the two lower sites (1050 and 1540 m) and Cambisols or Podzols found at the three upper ones (1890–3060 m). The organic layer atop of the mineral soil increases in thickness from about 48 mm at 1050 m to ca. 430 mm at 3060 m. There is a decrease in soil pH (H_2O) from 4.7 to 3.2 and an increase in C/N ratio (17–26) along the slope (mineral topsoil: 0–10 cm). Particularly large C/N ratios (46–63) were found in the uppermost organic horizons of the high elevation stands 4 and 5 (Iost et al., unpubl. data). In general, the soils of the study region are of low fertility with small cation exchange capacities and low base saturations (Schrumpf et al. 2001).

Analysis of stand structure. Tree biometric data were investigated in populations of each 80 trees per stand that covered 827 m^2, 360 m^2, 343 m^2, 290 m^2 and 96 m^2 (in horizontal projection) of the stands 1, 2, 3, 4 and 5, respectively. Thus, the size of the inventory plots decreased upslope due to increasing stem density. All trees reaching the canopy were investigated for stem length and diameter at breast height (DBH, at 1.3 m). Stem length and tree height were measured independently because many trees did not grow in erect position. We did not exclude trees of a certain DBH if they reached the canopy. Therefore, at 3060 m elevation, trees with a DBH as small as 3 cm were included, while 5 cm was the minimum DBH at 1050 m a.s.l.

Tree height was measured with a Vertex III Forestor tree height meter (Haglöf, Långsele, Sweden). The angle of the stems was also recorded for calculating total stem length in the steeply sloped stands; this data was then used for stem biomass calculation.

After projecting the 80-tree inventory plots on the horizontal plane, we calculated stem density and basal area, the latter being the sum of the cross-sectional areas of all trees derived from DBH.

We determined annual leaf litter production with 12 litter traps per plot using buckets equipped with 50 x 50 cm sheets of gauze with 1 mm mesh width, positioned randomly within the 20 x 20 m plots. The litter traps were emptied during the 1-year measuring period every three weeks in the stands #1 and 2, where decomposition rates were high, and every six weeks in the three other stands, starting on May 26, 2003. In the litter analysis, each two subsequent 3-week periods were added in the stands #1 and 2 to reach at 6-week periods which could be compared to the data of the other three stands. For estimating stand leaf biomass, annual leaf litter production and LAI, only the tree leaf fraction was considered. The small fractions of epiphyte leaves, leaves of understorey plants such as bamboo, and other components (e.g. flowers and fruits) were determined separately but not included in the analysis here.

All leaves collected in the litter traps (except for those from stand #5) were scanned at 300 dpi with a flat bed scanner at ECSF, and the leaf area was determined with the WinFolia 2000a program (Régent Instruments, Quebec, Canada). Specific leaf area (SLA) was calculated by relating leaf area to dry weight (48 h, 70° C). Since SLA showed no significant differences between the different sampling dates at any study site, we used the annual mean SLA value of a site for calculating leaf area from dry mass throughout the year. A somewhat different sampling protocol was developed for study site #5 where *Weinmannia loxensis* Harling (Cunoniaceae) was the dominant tree species. This species has small pinnate leaves with the leaflets being shed individually. SLA and leaf dry mass were determined separately for this species and the data included in the stand mean by its relative mass proportion.

We assumed that leaf growth and leaf abscission occurred at similar rates in the study year, i.e. that leaf biomass was equal in May 2003 and in May 2004. This would imply that annual leaf litter production (M_l) equalled annual leaf biomass production.

Leaf lifespan was recorded by establishing leaf survivorship curves for leaf populations of 10–15 trees per stand that were selected randomly in the plots. In all cases, small individuals (1.5 to 5 m high) of canopy tree species were selected since continuous access to the upper crown of large trees could not be provided. In the stands #1 to 4, these trees were part of the second and lower canopy strata, whereas they formed the uppermost canopy layer in stand #5. In May 2003, 24 twigs per plot were marked in these tree individuals. The total number of observed leaves ranged between 254 and 666 per stand. Every three to six weeks, the number of young, mature and dead leaves per twig marked was counted. The declining number of surviving leaves from the first census was then followed in leaf survivorship curves until the last leaf of this census was shed. In the stands #3 to 5, the average lifespan of many leaves exceeded the observation period of 19 months. In these cases, the average leaf lifespan of the leaf population was obtained from linear extrapolation of the survivorship curves.

Leaf area index and stand leaf biomass. In all five stands, we measured leaf area index (LAI) with a LAI-2000 canopy analyser system (Licor, Lincoln, Nebraska, USA) and additionally took digital hemispherical photographs for subsequent analysis with HemiView 2.1 software (Delta-T Devices, Cambridge, UK). The reference measurements with the LAI-2000 system were conducted either simultaneously over a grassland patch at a distance to the forest of 200–300 m (stands #1 and 2), or were taken immediately before and after the forest measurements in an open area within a 50 m-distance from the forest (stands #3 to 5).

All LAI-2000 measurements (n = 10 per stand) and hemispherical photogaphs (n = 12) were taken either before sunrise, after sunset or during uniform overcast sky conditions during daytime hours. The measurement locations were in direct vicinity of the litter traps at 1 m height. We took care that the canopy was dry during the LAI-2000 measurements and while taking the hemispherical photographs; otherwise, reflected light on wet surfaces would have led to an underestimation of LAI. The LAI-2000 sensor was positioned precisely in the horizontal plane during the measurements. In the analysis of the LAI-2000 data, we used only the three upper rings (0–43° from zenith) of the hemispherical sensor, as recommended by Dufrêne & Bréda (1995) for obtaining optimal LAI estimates with this system. For hemispherical photography, a Nikon Coolpix digital camera with a 180° fisheye lens was used. The camera was fixed in a self-levelling camera mount (type SLM5, Delta-T Devices) for guaranteeing horizontal orientation. The SLM5 system also provided markers for the horizon and north-south axis for assisting with image alignment.

To calculate the stand leaf biomass (B_l in g m^{-2}) of the study plots, two different equations were used:

$$(1) \qquad B_l = LAI\ SLA^{-1}$$

$$(2) \qquad B_l = M_l \cdot D_l$$

with M_l being the annual leaf litter production (in g m^{-2} yr^{-1}) and D_l the average leaf lifespan (in yr). For solving equation (1), the LAI values derived from the LAI-2000 measurements were used. SLA in equation (1) refers to the measured stand mean of SLA. We combined our annual leaf litter data with the results on average SLA and mean leaf lifespan to calculate LAI independently from the two optical approaches of leaf area estimation:

$$(3) \qquad LAI = M_l\ SLA\ D_l$$

Thus, stand leaf biomass was calculated with two, LAI with three independent approaches which allowed a comparison of optical and leaf mass-related techniques in the five forest stands.

Estimation of aboveground biomass. Aboveground tree biomass was estimated with allometric equations for the 80 canopy trees per plot based on the measured DBH and tree height data. We ignored understorey trees and shrubs, and standing or lying dead trunks (see Wilcke et al. 2005) since understorey biomass in mature moist tropical forests may comprise less than 3% of the aboveground biomass (Brown 1997).

We screened the literature for allometric equations available for humid tropical moist forests. In 38 relevant studies we found 184 different allometric equations for total aboveground tree biomass or different fractions of it, not including lianas (see Gerwing et al 2006, Schnitzler et al. 2006). The only allometric equation that seems to exist for Andean mountain forests was developed for a cloud forest in Venezuela (Brun 1976) with high specifity to the local conditions in that forest. Since specific equations applicable for lower to upper tropical montane forests have only been established for Malaysia (Yamakura et al. 1986), Jamaica (Tanner 1980) and Hawaii (Raich et al. 1997), we tested the pantropical equations established by Brown & Iverson (1992) and Chave et al. (2005) for estimating aboveground tree biomass only.

The pantropical equation of Brown & Iverson (1992) is based on DBH and tree height data; that of Chave et al. (2005) additionally considers wood specific density which can significantly improve the accuracy of biomass estimation from allometric regressions (Chave et al. 2004). These pantropical equations may be applicable to the whole range of Andean mountain forest types from the lower to the upper montane vegetation belt. The equations were developed for estimating the total aboveground biomass including leaves, twigs, branches, bark and boles of trees, based on data of 169 (Brown & Iverson 1992) or 2410 (Chave et al. 2005) harvested trees from all over the tropics.

The applicability of the allometric equations to our stands was assessed with three recently wind-thrown tree individuals per study site, which were analyzed for stem length, DBH, wood volume and specific wood gravity. Based on best fit to these empirical data we selected the allometric equation of Chave et al. (2005):

$$AGB = \exp(-2.557 + 0.940 \ln(\sigma D^2 H))$$

where *AGB* is tree aboveground biomass (kg per tree), *D* is DBH (cm), *H* is stem height (m) and σ is wood density (g cm^{-3}). The stand aboveground biomass total (AGB) was obtained by summing up the calculated masses of the 80 trees per plot.

Root sampling and analysis. Coarse root biomass for each stand was determined in 12–16 soil pits (40×40 cm) that were dug to 60 cm soil depth (data for site #3-5 provided by N. Soethe, Berlin). Biomass (live roots) and necromass (dead roots) of all roots with a diameter > 2 mm were excavated in steps of 10 cm horizons in the organic layer and the mineral soil. In the laboratory, all roots were washed and dried at 70 °C to constant dry mass.

For analysing the bio- and necromass of fine roots (diameter < 2 mm), soil coring was conducted from March to May 2003 in soil profiles of 30 cm depth (organic layer and mineral soil) under the five stands. Preliminary investigation of a limited number of soil cores to 60–80 cm depth revealed that the organic layer and the mineral soil to 30 cm depth must contain about 75% or more of the profile total of tree fine root biomass since fine root densities in the subsoil were very low. Profile totals of root biomass investigated in this study refer to the organic layer and the mineral soil to 30 cm depth.

Root sampling was conducted with a steel corer (33 mm in diameter, n = 20 per plot). The soil material was stored at 4 °C in the laboratory at Estación Científica San Francisco where processing took place within 30 days. Fine root biomass and necromass were separated under a microscope according to the procedure described by Leuschner et al. (2001).

Results

Canopy height declined 3.5-fold and mean stem length 3.0-fold between 1050 m and 3060 m (Table 1). In the uppermost stand (plot #5) in the elfin forest, strongly inclinated trunks with low canopy height were responsible for the relatively high maximum stem length (Table 1). We found an 8.6-fold decrease in maximum DBH and a 2.5-fold one in mean DBH between plot #1 and plot #5 which is roughly proportional to the 3.0-fold reduction in mean stem length, but contrasts with the 8.6-fold increase of stem density along the 2000-m elevation transect. In contrast, stand basal area revealed no elevational change.

Table 1. Aboveground stand structural characteristics of the five study plots.

plot	elevation [m]	canopy height [m]	stem length [m]	dbh [cm]	stem density [n ha^{-1}]	basal area [m^2 ha^{-1}]
1	1050	31.8	15.6 ±0.7[a]	17.3±1.3[a]	968	33.6
2	1540	21.7	12.1±0.5[b]	11.5±0.6[b]	2222	27.5
3	1890	18.9	10.1±0.4[c]	12.2±0.8[b]	2333	36.9
4	2380	12.0	7.4±0.3[d]	9.8±0.6[c]	2753	27.2
5	3060	9.0	5.2±0.3[e]	7.2±0.4[d]	8317	42.2

Total aboveground tree biomass (AGB) declined 2.5-fold from plot #1 to plot #5 and was significantly dependent on elevation (Table 2). In plot #1, trees with a DBH > 30 cm accounted for more than two-third of total aboveground tree biomass. In contrast, trees with a DBH of 10–20 cm were responsible for the largest portion of AGB in all other stands.

Table 2. Aboveground (AGB), belowground (BGB) and total biomass, and root/shoot ratio of the five stands in South Ecuador

plot	elevation [m]	AGB [Mg ha^{-1}]						BGB [Mg ha^{-1}]	total biomass [Mg ha^{-1}]	root/ shoot
		3-5 cm	5-10 cm	10-20 cm	20-30 cm	30-70 cm	total[a]			
1	1050		7.7	40.9	41.1	195.3	285.1	32.1	317.2	0.113
2	1540	0.4	19.8	97.3	50.0		167.5	36.3	203.8	0.217
3	1890	0.4	22.4	119.4	30.8		173.0	25.8	198.8	0.149
4	2380	0.8	19.1	49.8	7.2	22.9	99.8	39.2	139.0	0.393
5	3060	8.8	51.1	52.3			112.2	62.7	174.9	0.559

[a] Estimates of total aboveground tree biomass are based on an allometric equation given by Chave et al. (2005), which relates biomass to tree height, DBH and wood density. Given are AGB values for five different DBH classes. Belowground biomass is given as the sum of coarse root biomass (diameter > 2 mm) and fine roots (diameter < 2 mm). Total biomass is the sum of AGB and BGB. The root/shoot biomass ratio was calculated as the BGB/AGB quotient.

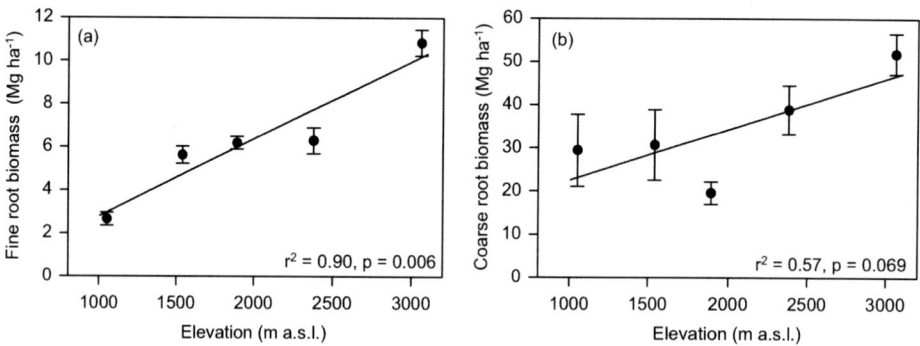

Figure 1. Elevational change of (**a**) fine root biomass (diameter < 2 mm, means and SE, n = 20) and (**b**) and coarse and large root biomass (n = 12). Coarse root data for site #3–5 by N. Soethe (Humboldt Univ. of Berlin).

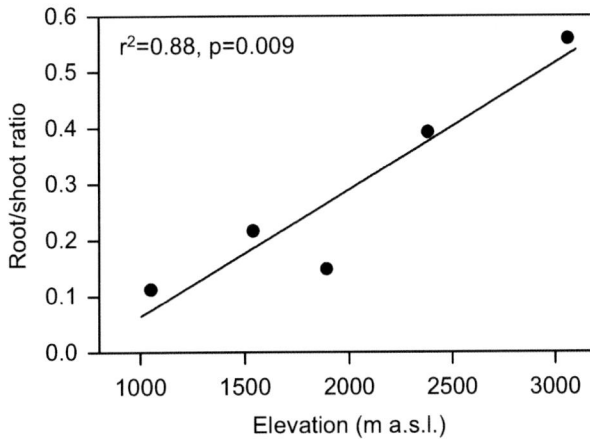

Figure 2. Linear regression analysis between the root/shoot biomass ratio, calculated as the quotient of belowground biomass (BGB) to aboveground biomass (AGB), and elevation.

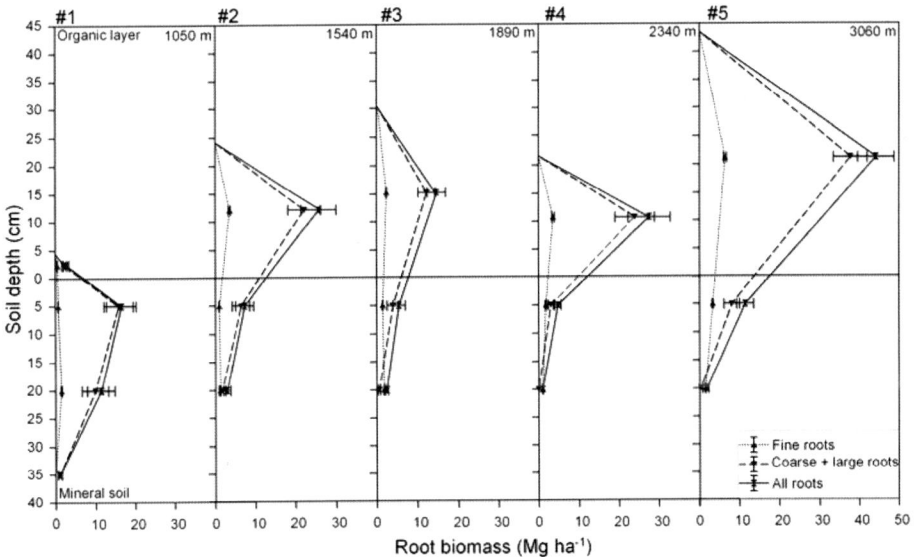

Figure 3. The vertical distribution of fine root, coarse and large root, and total root biomass in the five Ecuadorian plots. The position of the data points on the vertical axis indicates the upper limit and the centre of the organic layer and the various mineral soil horizons (0–10, 10–30 and 30–40 cm), respectively. Given are means $\pm$ 1 SE of root biomass (n=15).

We also found a large and highly significant increase in fine root necromass (FRN) with elevation (1.67–31.06 Mg ha⁻¹). The biomass of coarse and large roots ranged between 0.17 and 0.23 Mg ha⁻¹ for sites #1, 3 and 4 and sites #2 and 5 with 0.79 and 0.89 Mg ha⁻¹, respectively (Fig. 1b). Coarse root necromass was comparably small at all sites with the highest value again found in the highest stand (data not shown).

The marked decrease in aboveground biomass (from 285 to 112 Mg ha-1) and the doubling in root biomass (from 32 to 63 Mg ha-1) resulted in a large increase of the root/shoot ratio, i.e. the BGB/AGB ratio (Fig. 2).

The portion of belowground biomass in total tree biomass increased from 10.1% to 35.8% between the lowermost and the uppermost plot. The five-fold increase in the root/shoot ratio of the trees between 1050 m and 3060 m in the Ecuadorian transect underlines the large carbon allocation shift from aboveground to belowground tree organs that takes place along this slope (Table 2).

Fig. 3 shows the vertical distribution of fine, coarse and total root biomass in the 5 stands. The absolute increase of root biomass with elevation is mainly caused by a steep increase of root biomass within the organic layer. Total root biomass in the organic layer was very low in forest stand #1 at 1050 m elevation, due to the limited depth of this horizon. The bulk of root biomass was concentrated in the organic layer in all other forest stands and reached 83% in plot #4 and 77% in plot #5. At the lowermost plot (#1), half of total root biomass was concentrated in the uppermost 10 cm of the mineral soil. Only very few roots were observed below 40 cm depth. In all other forest stands, the decline of root biomass from the organic layer to the mineral topsoil horizon was very steep and only very few roots occurred below 30 cm depth. Forest plots #2 (1540 m) and #4 (2340 m) showed very similar vertical root distribution patterns; both stands had a higher root biomass in the organic layer than in plot #3.

Table 3. Annual leaf litter production as measured with litter traps and stand averages of specific leaf area (SLA) and leaf lifespan in the five stands of the Ecuador transect.

plot	1	2	3	4	5
leaf litter production [g m⁻² yr⁻¹]	505 ± 33[a]	502 ± 35[a]	496 ± 43[a]	266 ± 22[b]	179 ± 15[c]
SLA [cm²g⁻¹]	87.9 ± 1.5[a]	67.8 ± 1.4[b]	58.5 ± 1.8[c]	52.7 ± 1.8[d]	61.1 ± 2.4[c]
leaf lifespan [months] (n)	16.2 ± 2.6[a] (306)	19.1 ± 3.8[ab] (316)	23.6 ± 2.3[c] (254)	23.7 ± 2.7[bc] (594)	24.5 ± 2.3[c] (666)

Given are means ± 1 SE and the number of leaves observed for lifespan. Different letters indicate significant differences between study sites (P < 0.05).

The stand average of specific leaf area (SLA) declined significantly from plot #1 to plot #4, but was higher again in the uppermost stand (Table 3). The apparent increase from stand #4 to #5 was primarily caused by the high SLA value of the dominant tree species *Weinmannia loxensis* in the uppermost stand, which is characterized by pinnate and thin leaves. Average leaf lifespan increased significantly from plot #1 to #3, and remained more or less constant further upslope (Table 3). No elevational trend of annual leaf litter production (M_l) was visible between the plots #1 and #3. Further upslope, however, M_l decreased to about a third in plot #5 (Table 3).

The three different approaches for estimating LAI showed only partial agreement along the transect. The LAI estimate obtained from the LAI-2000 measurements decreased linearly (Table 4) with all differences between the stands being significant, except for the plots #3 and 4 which had similar LAI-2000 values. LAI estimation with hemispherical photography gave, in general, lower values than the LAI-2000 measurement. The discrepancy between the two optical methods was particularly large at lower elevation (< 2000 m). Hemispherical photography yielded no significant differences in LAI between any of the plots #1 to 4 (Table 4). Finally, the results from the mass-related calculation, based on leaf litter production, leaf lifespan and SLA, were much closer to the LAI-2000 data than those from hemispherical photography. In agreement with this observation, a significant dependence on elevation was only found for these two approaches but not for hemispherical photography. The leaf mass-related approach resulted in similarly high values for the plots #1 to 3 (differences not significant) and showed a steep decrease further upslope (Table 4).

Table 4. Leaf area index (LAI) estimates for the five stands of the Ecuador transect as obtained with three different approaches.

plot	1	2	3	4	5
LAI [$m^2\,m^{-2}$]					
LAI-2000	5.1 ± 0.1^{aA}	4.6 ± 0.1^{bA}	3.9 ± 0.2^{cA}	3.6 ± 0.1^{cA}	2.9 ± 0.3^{dA}
hemispherical photography	2.8 ± 0.2^{aB}	3.0 ± 0.3^{aB}	3.0 ± 0.3^{aA}	2.7 ± 0.1^{aB}	2.2 ± 0.2^{bA}
leaf litter production + leaf lifespan + SLA data	6.0 ± 0.4^{aA}	5.4 ± 0.4^{aA}	5.7 ± 0.5^{aB}	2.8 ± 0.2^{bB}	2.2 ± 0.2^{cA}

Given are means ± 1 SE; different small letters indicate significant differences between the five stands, different capitals indicate significant differences between the three approaches for a given stand.

The three approaches for estimating LAI agreed only in the uppermost stand (#5). Besides the very low values obtained by hemispherical photography on the lower slope, there was a general tendency of the mass-related approach for yielding highest values at lower elevations (< 2000 m) and a tendency of the LAI-2000 method for doing so at the upper elevations (> 2000 m).

Discussion

Altitudinal change of stand structure. A consistent pattern of structural changes with increasing elevation can be observed: Tree height or stem length, DBH and leaf area index show a general decrease with elevation whereas stem density increases with altitude. In the large-scale altitudinal transects in Malaysia (Aiba & Kitayama 1999), Costa Rica (Lieberman et al. 1996) and Hawaii (Raich et al. 1997), tree height showed a more or less continuous decrease with altitude like in our Ecuadorian transect. Similarly, a continuous decrease of tree height from the lowlands to the timberline was also reported for undisturbed temperate mountain forests, e.g., in the Southern Alps of Italy (Reisigl & Keller 1999) and in Tierra del Fuego (Pollmann & Hildebrand 2005). However, in certain mountain transects as on Tenerife, tree height seems to decrease upslope only slightly; the timberline may then consist of tall trees (Srutek et al. 2002). We speculate that a more or less continuous decrease in tree height occurs on mountain slopes where the altitudinal temperature decrease is the main environmental factor that controls tree growth and microbial activity in the soil. However, in all those mountains, where additional environmental constraints such as waterlogging, drought or strong winds are influencing tree growth, these factors may overlay the temperature effect causing a more or less abrupt transition between tall high-elevation trees and low-statured krummholz or alpine non-forest vegetation higher upslope. An important second cause of abrupt timberlines is human impact that has lowered timberline elevation in many mountains of the tropics.

Altitudinal comparisons of mean DBH are often problematic because authors tend to set lower limits to stem diameter measurements at higher elevation where thinner stems prevail than in low-elevation forests with thicker stems. When the lower DBH limit varies along the slope, a larger altitudinal decrease in mean DBH may be detected than exists in reality. Liebermann et al. (1996) used a 10 cm DBH minimum in their Costa Rican elevation gradient between 100 and 2600 m and did not detect altitudinal effects but found highest means at high elevations. To avoid such a bias in the Ecuadorian transect, we have investigated all stems that reach the canopy, irrespective of diameter. We found a decrease in mean DBH by a factor of 2.5 between stand #1 and #5 which is roughly proportional to the reduction in mean stem length from 15.6 to 5.2 m.

Smaller trees with thinner stems and less extended crowns allow for higher tree densities per ground area when moving upslope in mountains. Altitudinal increases in tree density were not only reported for South Ecuador, but also for a transect on Mt. Kinabalu in Malaysia (Takyu et al. 2002). In contrast, Heaney & Proctor (1990) found only minor changes in stem density between 100 and 2600 m on Volcán Barva, Costa Rica, probably because they also used a minimum DBH of 10 cm at all elevations. Similarly, upper-montane *Quercus* forests in the Sierra de Talamanca (Costa Rica) at 2900 m had exceptionally low stem densities when only stems with a DBH greater 10 cm were considered (390 ha^{-1}). These values are not higher than in many tropical lowland forests. However, if all stems > 3 cm DBH are included, stem density increases by a factor of nearly ten (3460 ha^{-1}; Köhler 2002). On subtropical Mt. Teide, Tenerife, the density of *Pinus canariensis* trees remains constant with elevation or decreases (Srutek et al. 2002). We conclude that changes in tree density along tropical mountain slopes are probably highly dependent on the floristic composition and the stand dynamics of the respective forest communities. In addition, tree density in high-elevation forests may also depend on local edaphic and climatic conditions such as the occurrence of waterlogging or exposure to strong winds. Temporarily waterlogged soils favor woody plants that are able to resprout and to form multiple stems giving them a high morphological plasticity. For example, the widespread species *Weinmannia loxensis* (Cunoniaceae) in the uppermost stand (# 5 at 3060 m) of the Ecuador transect forms creeping belowground stems which connect up to three shoots (Soethe et al. 2006). The ability of many tree species to grow multiple stems on temporarily waterlogged soils partly explains the very high stem densities that were counted in stand #5 (8317 ha^{-1}).

The values of Madsen & Øllgaard (1994) (2568 and 3583 ha^{-1}) from upper montane forests of this region (2700 and 2900 m elevation) are lower than this value but they also indicate high tree densities close to the timberline.

Stand basal area is a function of mean DBH and stem density which both showed opposite trends with elevation in South Ecuador. In this transect, we found no clear altitudinal dependence of basal area although the highest value was measured at the uppermost stand. This is in line with results obtained by Lieberman et al. (1996) in Costa Rica. These results together with literature data indicate that basal area of tropical mountain forests seems to be only weakly dependent on climatic, edaphic or stand structural parameters such as maximum tree height (Aiba & Kitayama 1999).

Altitudinal change of stand leaf biomass and leaf area index. Our estimates of stand leaf biomass (B$_l$) at lower to mid-montane elevation (stands #1-3: 6.8-9.7 Mg ha^{-1} according to the leaf mass-related approach) are comparable to the mean B$_l$ value given by Raich et al. (1997) for tropical montane forests (8.1 Mg ha^{-1}). However, our data are also in the range of leaf biomass values reported from tropical lowland forests (6.3-10.8 Mg ha^{-1}, McWilliam et al. 1993) indicating that the difference in B$_l$ between lowland and montane stands is small. The low B$_l$ value from the uppermost stand in Ecuador points to a leaf mass decrease only in the uppermost section of the transect. This conclusion is supported by the regression analysis of stand leaf biomass on elevation which revealed no significant relationship neither for the leaf mass-related nor the LAI-2000 estimate. This result contrasts with a decreasing leaf biomass along an elevation transect (290–1660 m) on young lava flows in Hawaii (Raich et al. 1997), and on ultrabasic bedrock on Mt. Kinabalu, Malaysia (Kitayama & Aiba 2002). However, a second transect on sedimentary soils on Mt. Kinabalu showed a slight increase of B_l with elevation supporting the view that the leaf biomass difference between lowland and montane forests is generally small (Tanner 1980).

According to Medina & Klinge (1983), the annual leaf litter production (M_l) of tropical montane forests is generally lower than that of tropical moist lowland forests (3.9–6.4 vs. 5.3–8.2 Mg ha^{-1} yr^{-1}). These results compare well with our data which showed a litter production of 5.0 Mg ha^{-1} yr^{-1} for the lower and mid-montane stands (#1-3) but a sharp decrease higher upslope at elevations > 2000 m (stands #4 and 5). In an elevational transect study in tropical moist forests in Hawaii, M_l decreased with elevation from 290 to 1660 m (Raich et al. 1997).

Although specific leaf area is highly dependent on species, the stand average of SLA tends to decrease with elevation in tropical mountains. For example, Medina & Klinge (1983) reported an SLA value of 83 cm^2 g^{-1} for a tropical montane forest in Puerto Rico compared to 47 cm^2 g^{-1} in an elfin forest higher upslope. In contrast, tropical lowland moist forests have average SLA values in the range of 65 to 93 cm^2 g^{-1} (McWilliam et al. 1993, Medina & Klinge 1983).

In the Ecuador transect, specific leaf area decreased from 88 cm^2 g^{-1} at 1050 m to 53 cm^2 g^{-1} at 2380 m while it seemed to increase again further upslope toward the uppermost stand (61 cm^2 g^{-1}). This picture changes if the high SLA value of the most abundant tree species, *Weinmannia loxensis* with pinnate leaves (87 cm^2 g^{-1}), that are atypical for this elevation, is omitted from the analyis: the mean SLA of stand #5 would then be much lower (57 cm^2 g^{-1}), supporting the general negative trend of SLA with elevation. Thus, despite a large species effect on leaf morphology, it is evident that SLA shows a much steeper decrease with elevation than annual leaf litter production (and annual leaf biomass production) or stand leaf biomass.

LAI decreased significantly with elevation in the Ecuador transect according to both the leaf mass-related and the LAI-2000 estimate. This observation compares well with the results of a pan-tropical meta-analysis from tropical montane forests by Moser et al. (2007) where the regression line based on 49 stands indicates a mean LAI of 4.9 m^2 m^{-2} at 1000 m and a decrease further upslope by about 1 m^2 m^{-2} km^{-1}. Since the majority of data points were obtained by LAI-2000 measurements, one may assume from the specific errors associated with this method that the LAI decrease with elevation is even steeper than indicated by the data.

A decrease in LAI must have important implications for the prediction of forest productivity along tropical mountain slopes. Since canopy carbon gain is closely linked to the amount of intercepted radiation, a marked reduction in LAI must cause a substantial decline in net primary production with increasing elevation, independently from any possible negative effect of lowered temperatures or nutrient shortage on canopy carbon gain.

Besides a reduced leaf area, a reduction in foliar nitrogen content may be another factor that could lead to a lowered photosynthetic carbon gain in high-elevation forests. In the Ecuador transect, the foliar C/N ratio greatly decreased from 1050 to 3060 m while, due to a substantial reduction in SLA, foliar N (nitrogen) content per unit leaf area did not (M. Unger, unpublished). Even though data on photosynthetic capacity are not available from this transect, it can be concluded that a 40% reduction in LAI must represent the principal factor controlling elevational changes in canopy carbon gain, whereas the decrease in foliar N content will be of secondary importance because it is more or less compensated by a larger leaf thickness.

Altitudinal change of tree biomass and root/shoot ratio. Total aboveground tree biomasses significantly decreased with altitude in the Ecuadorian transect. Temperature and C/N ratio of the organic layer are variables that also showed a similarly close relation to AGB as did elevation because both are strongly correlated with elevation. A causal explanation of the reduction in AGB must remain speculative since elevation, temperature, soil C/N ratio and also soil pH are closely related to each other in the Ecuadorian transect. Specific hypotheses require experimental testing.

A striking result of this study is the high fine and coarse root biomass of the high-elevation forests in Ecuador. Belowground biomass (BGB) of the uppermost stand (# 5 at 3060 m) was about two times greater than that of the stands # 1 to 3 (1050 to 1890 m). Corresponding trends in fine root biomass with elevation have been reported by Kitayama & Aiba (2002) on Mt. Kinabalu, and in a meta-analysis of fine root data of paleo- and neotropical forests by Hertel & Leuschner (2008). Thus, a direct effect of lowered temperature on aboveground meristematic activity and tree growth is unlikely because air and soil temperature differed by not more than 1 °C from each other, and branch and root meristems should have responded similarly to a reduction in temperature along the slope. Rather, a direct or indirect effect of lowered temperature on carbon allocation patterns of the trees must play a prominent role in the altitudinal reduction in AGB.

A fivefold increase in the root/shoot ratio of the trees between 1050 and 3060 m in the Ecuador transect underlines the large carbon allocation shift from aboveground to belowground tree organs that takes place along this slope.

Conclusions

According to the resource balance hypothesis of Bloom et al. (1985) the impressive change in allocation patterns is best explained by a growing importance of limiting soil resources over limiting light with increasing elevation. Increasing carbon and nutrient allocation to roots on the cost of aboveground biomass would then represent a compensatory response of the trees to cope with an increasing limitation by nutrient and/or water shortage at high elevations.

Water shortage is unlikely to occur at the high-elevation sites in South Ecuador; rather, water logging may have contributed to a slowing down of decomposition and thus a poor nutrient supply in the uppermost stands (Schuur 2001). There are several possible pathways by which low temperatures may have resulted in impaired nutrient supply or uptake, among them: (a) reduction in decomposition rate, (b) low activity of mycorrhizal fungi and (c) reduction in membrane transporter activity and thus lowered nutrient uptake rate.

Unfavorable soil chemical conditions as a consequence of low nutrient contents in plant litter, or high concentrations of free aluminum and elevated Al/Ca ratios in the soil solution (Hafkenscheid 2000) may also negatively affect root vitality and root growth, additionally impairing nutrient uptake. Finally, the observed large stocks of coarse root biomass in the uppermost stands, which mainly serve for tree anchoring, point at the growing need for tree stabilization on steep and wet slopes at high elevations (Soethe et al. 2006).

While it is unlikely that a single factor is responsible for the observed remarkable carbon allocation shift with increasing elevation, there is an urgent need of well designed field experiments aimed at disentangling the possible influencing factors that may limit nutrient supply and nutrient uptake in these high-elevation tropical forests.

Acknowledgements. This research was conducted in the framework of Research Unit 402 ("Funktionalität in einem tropischen Bergregenwald Südecuadors", subproject B7) funded by the German Research Foundation (DFG). Financial support is gratefully acknowledged. We are also grateful to Marina Röderstein and Jürgen Homeier, who provided part of the data presented here.

References

Aiba S, Kitayama K (1999) Structure, composition and species diversity in an altitude-substrate matrix of rain forest tree communities on Mount Kinabalu, Borneo. Plant Ecology 140: 139-157

Ashton PS (2003) Floristic zonation of tree communities on wet tropical mountains revisited. Perspectives in Plant Ecology Evolution and Systematics 6: 87-104

Balslev, H, Øllgaard B (2002) Mapa de vegetacion del sur de Ecuador In: Aguirre M, Madsen JE, Cotton E, Balslev H (eds), Botanica Austroecuadoriana. Estudios sobre los recursos vegetales en las provincias de El Oro, Loja y Zamora-Chinchipe. Ediciones Abya-Yala, Quito, Ecuador, pp 51-64

Bloom AJ, Chapin III FS, Mooney HA (1985) Resource limitation in plants - an economic analogy. Annual Review of Ecology and Systematics 16: 363-392

Brown S (1997) Estimating biomass and biomass change of tropical forests: a primer. FAO Forestry Paper 134

Brown S, Iverson LR (1992) Biomass estimates for tropical forests. World Resource Review 4: 366-384

Bruijnzeel LA, Proctor J (1995) Hydrology and biogeochemistry of tropical montane cloud forest: What do we really know ? In: Hamilton LS, Juvik JO & Scatena FN (eds) Tropical montane cloud forests. Springer, New York. Ecological Studies 110, pp 38-78

Bruijnzeel LA, Veneklaas EJ (1998) Climatic conditions and tropial montane forest productivity: the fog has not lifted yet. Ecology 79: 3-9

Brun R (1976) Methodik und Ergebnisse zur Biomassenbestimmung eines Nebelwald-Ökosystems in den venezolanischen Anden. Proc. 16th IUFRO World Congress, Div. I, Oslo, Norway: 490-499

Cairns MA, Brown S, Helmer EH, Baumgardner GA (1997) Root biomass allocation in the world's upland forests. Oecologia 111: 1-11

Chave J, Andalo C, Brown S, Cairns MA, Chambers JQ, Eamus D, Fölster D, Fromard F, Higuchi N, Kira T, Lescure J-P, Nelson BW, Ogawa H, Puig H, Riéra B, Yamakura T (2005) Tree allometry and improved estimation of carbon stocks and balance in tropical forests. Oecologia 145: 87-99

Chave J, Condit R, Aguilar S, Hernandez A, Lao S, Perez R (2004) Error propagation and scaling for tropical forest biomass estimates. Philosophical Transaction of the Royal Society London, Ser B 359: 409-420

Dufrêne E, Bréda N (1995) Estimation of deciduous forest leaf-area index using direct and indirect methods. Oecologia 104: 156-162

Flenley JR (1995) Cloud forest, the Massenerhebung effect, and ultraviolet insolation In: Hamilton LS, Juvik JO, Scatena FN (eds) Tropical montane cloud forests. Springer, New York. Ecological Studies 110, pp 150-155

Gerwing JJ, Schnitzer SA, Burnham RJ, Bongers F, Chave J, DeWalt SJ, Ewango CEN, Foster R, Kenfack D, Martinez-Ramos M, Parren M, Parthasarathy N, Perez-Salicrup DR, Putz FE & Thomas DW (2006) A standard protocol for liana censuses. Biotropica 38: 256-261

Grubb PJ (1977) Control of forest growth and distribution on wet tropical mountains: with special reference to mineral nutrition. Annual Review of Ecology and Systematics 8: 83-107

Hafkenscheid RLLJ (2000) Hydrology and biogeochemistry of tropical montane rain forests of contrasting stature in the Blue Mountains, Jamaica. Dissertation, University of Amsterdam. Print Partners Ipskamp, Enschede, The Netherlands

Heaney A, Procter J (1990) Preliminary studies on forest structure and floristics on Volcán Barva, Costa Rica. Journal of Tropical Ecology 6: 307-320

Hertel D, Leuschner Ch (2008) Fine root mass and fine root production in tropical moist forests as dependent on soil, climate and elevation In: Bruijnzeel LE, Juvik JO (eds) Mountains in the mist: science for conserving and managing tropical montane cloud forests. Hawaii University Press, in press

Hertel D, Leuschner Ch, Hölscher D (2003) Size and structure of fine root systems in old-growth and secondary tropical montane forests (Costa Rica). Biotropica 35: 143-153

Hetsch W, Hoheisel K (1976) Standorts- und Vegetationsgliederung in einem tropischen Nebelwald. Allgemeine Forst- und Jagdzeitung 147: 200-209

Homeier J (2004) Baumdiversität, Waldstruktur und Wachstumsdynamik zweier tropischer Bergregenwäldern in Ecuador und Costa Rica. Dissertationes Botanicae 391: 1-207

Kitayama K & Aiba S-I (2002) Ecosystem structure and productivity of tropical rain forests along altitudinal gradients with contrasting soil phosphorus pools on Mount Kinabalu, Borneo. Journal of Ecology 90: 37-51

Köhler L (2002) Die Bedeutung der Epiphyten im ökosystemaren Wasser- und Nährstoffumsatz verschiedener Altersstadien eines Bergregenwaldes in Costa Rica. Dissertation, University of Göttingen. Berichte des Forschungszentrums Waldökosysteme, Reihe A, Bd. 181: 1-134

Leuschner Ch, Hertel D, Coners H, Büttner V (2001) Root competition between beech and oak: a hypothesis. Oecologia 126: 276-284

Lieberman D, Lieberman M, Peralta R, Hartshorn GS (1996) Tropical forest structure and composition on a large-scale altitudinal gradient in Costa Rica. Journal of Ecology 84: 137-152

Madsen JE, Øllgaard B (1994) Floristic composition, structure, and dynamics of an upper montane rain-forest in Southern Ecuador. Nordic Journal of Botany 14: 403-423

McWilliam ALC, Roberts JM, Cabral OMR, Leitao MVBR, Decosta ACL, Maitelli GT, Zamparoni CAGP (1993) Leaf-area index and aboveground biomass of terra-firme rain-forest and adjacent clearings in Amazonia. Functional Ecology 7: 310-317

Medina E, Klinge H 1983 Productivity of tropical forests and tropical woodlands In: Lange et al (eds) Physiological Plant Ecology - Encyclopedia of plant physiology, Vol. 12d. Springer, New York, pp 281-303

Moser G, Hertel D, Leuschner C (2007) Altitudinal change in LAI and stand leaf biomass in tropical montane forests: a transect study in Ecuador and a pan-tropical meta-analysis. Ecosystems 10: 924-935

Pollmann W, Hildebrand R (2005) Structure and the composition of species in timberline ecotones of the Southern Andes In: Broll G, Keplin B (eds) Mountain ecosystems - studies in treeline ecology. Springer, Berlin, pp 117-151

Raich JW, Russell AE, Vitousek PM (1997) Primary productivity and ecosystem development along an elevational gradient on Mauna Loa, Hawai'i. Ecology 78: 707-721

Reisigl H, Keller R (1999) Lebensraum Bergwald. Spektrum Akademie Verlag, Heidelberg.

Schnitzer SA, DeWalt SJ, Chave J (2006) Censusing and measuring lianas: A quantitative comparison of the common methods. Biotropica 38: 581-591

Schrumpf M, Guggenberger G, Valarezo C, Zech W (2001) Tropical montane rain forest soils – Development and nutrient status along an altitudinal gradient in the South Ecuadorian Andes. Die Erde 132: 43-59

Schuur EAG (2001) The effect of water on decomposition dynamics in mesic to wet Hawaiian montane forests. Ecosystems 4: 259-273

Soethe N, Lehmann J, Engels C (2006) Root morphology and anchorage of six native tree species from a tropical montane forest and an elfin forest in Ecuador. Plant and Soil 279: 173-185

Srutek M, Dolezal J, Hara T (2002) Spatial structure and associations in a Pinus canariensis population at the treeline, Pico del Teide, Tenerife, Canary Islands. Arctic, Antarctic and Alpine Research 34: 201-210

Sugden AM (1986) The montane vegetation and flora of Margarita Island, Venezuela. Journal of the Arnold Arboretum 67: 187-232

Takyu M, Aiba S-I, Kitayama K (2002) Effects of topography on tropical lower montane forests under different geological conditions on Mount Kinabalu, Borneo. Plant Ecology 159: 35-49

Tanner EVJ (1980) Studies on the biomass and productivity in a series of montane rain forests in Jamaica. Journal of Ecology 68: 573-588

Tanner EVJ, Vitousek PM, Cuevas E (1998) Experimental investigation of nutrient limitation of forest growth on wet tropical mountains. Ecology 79: 10-22

Valencia R, Ceron C, Palacios W, Sierra R (1999) Las formaciones naturales de la sierra del Ecuador In: R Sierra (ed) Propuesta preliminar de un sistema de clasificacion de vegetacion para el Ecuador continental. Proyecto INEFAN/GEF-BIRF y EcoCiencia, Quito, pp 79-108

van Steenis CGGJ (1972) The Mountain Flora of Java. Brill, Leiden

Vogt KA, Vogt DJ, Palmiotto PA, Boon P, O'Hara J, Asbjornson H (1996) Review of root dynamics in forest ecosystems grouped by climate, climatic forest type and species. Plant and Soil 187: 159-219

Whitmore TC (1998) An introduction to tropical rain forests, 2nd ed. Oxford University Press, Oxford

Wilcke W, Yasin S, Schmitt A, Valarezo C, Zech W (2008) Soils. In: Beck E, Bendix J, Kottke I, Makeschin F, Mosandl R (eds) Gradients in a tropical mountain ecosystem of Ecuador. Ecological Studies 198. Springer, Berlin, Heidelberg, New York, pp 75-85

Yamakura T, Hagihara A, Sukardjo S, Ogawa H (1986) Aboveground biomass of tropical rain-forest stands in Indonesian Borneo. Vegetatio 68: 71-82

Biodiversity and Ecology Series (2008) 2: 129-136
The Tropical Mountain Forest – Patterns and Processes in a Biodiversity Hotspot
edited by S.R. Gradstein, J. Homeier and D. Gansert
Göttingen Centre for Biodiversity and Ecology

Hydrology of natural and anthropogenically altered tropical montane rainforests with special reference to rainfall interception

Dirk Hölscher

Tropical Silviculture and Forest Ecology, Burckhardt-Institute, University of Göttingen, Büsgenweg 1, 37077 Göttingen, Germany, dhoelsc@uni-goettingen.de

Abstract. The water budget of tropical montane rainforests under near-natural conditions is characterized by high rainfall, the possibility of additional water input from cloud interception, low transpiration rates, high streamflow and high rainfall interception losses. A montane forest in the Ecuadorian Andes received 2504 mm yr^{-1} rainfall, from which 41% went into stream flow, 40% into rainfall interception and 19% into transpiration. Forest use or disturbance by logging leads to changes in forest structure, including decrease in tree height. In Sulawesi, Indonesia, rainfall interception diminished significantly along a gradient of forest use intensity (from 30% on average in natural forest to 18–20% in used forest stands) and was strongly correlated with tree height ($r^2 = 0.63$). By means of satellite image analyses and modelling it was possible to predict rainfall interception from canopy reflectance characteristics of these forest stands. Application of the model at the landscape level (784 ha) provided realistic results with high interception values being rare and restricted to natural forest stands distant to villages, whereas low interception characterized the intensively used sites close to settlements. I conclude that 1) rainfall interception is a major component of the water budget in montane tropical rainforests, 2) forest use can significantly reduce rainfall interception, and 3) satellite image analysis can successfully be applied for predicting interception at the regional scale.

Introduction

Mountain forests in the humid tropics can provide important water-related ecosystem services. They may serve as source areas of water for the generation of hydropower and for providing drinking water, and may offer partial protection against flash floods. Some tracts of tropical mountain forests (TMF) still remain relatively undisturbed, but many are under severe threat, are disturbed or have been converted to other land cover. Forest conversion to agricultural land cover types is well known to alter ecosystem hydrological fluxes significantly, with often adverse effects on ecological services (Bonnell & Bruijnzeel 2005). Comparatively little information is available, however, on the impact on hydrological fluxes of gradual changes in vegetation structure, as created by different forest use practices. Such gradual changes may e.g. be the result of buffer zone management around protected areas or of selective logging activities. There is increasing awareness that large forest areas are encroached by logging. For example, selective logging doubled previous estimates of the total annually amount of forest degraded by human activities in five states of the Brazilian Amazon lowlands (Asner et al. 2005). The aims of the present chapter on the hydrology of natural and altered TMFs are: (i) to report on recent water budget studies, (ii) to analyze forest stand structural characteristics presumed to be important for specific components of the balance such as rainfall interception, and (iii) to compare hydrological fluxes among natural and anthropogenically altered forests.

Water budget

In the Ecuadorian Andes, the water budget of undisturbed TMF was determined in catchment areas between 1900 to 2150 m (Fleischbein et al. 2006). Incident precipitation was 2504 mm yr[-1] and fog water input was considered negligible. Surface (stream) flow was 1039 mm per year and annual evapotranspiration 1466 mm; 32% of the evapotranspiration was transpiration and 68% interception loss (Fig. 1).

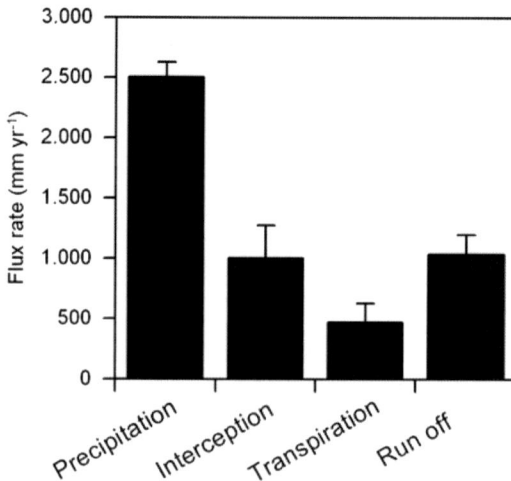

Figure 1. Water budget of a tropical mountain forest in near-natural condition in the Ecuadorian Andes (Fleischbein et al. 2006). The results are based on measurements in three catchment areas during three years.

Cloud water interception

In addition to conventionally measured rainfall, input from intercepted fog and cloud water may sometimes be quite high in TMF. By using the standard method of throughfall, stemflow and gross precipitation measurement, the additional input can only be computed for periods without rainfall. These measurements may in many cases represent underestimations, especially where low clouds and fog persist. For the Sierra de las Minas in Guatemala, an equivalent of fog precipitation of 203 mm yr^{-1} was estimated by applying this approach (Holder 2003). In three montane rainforests of Australia, cloud interception constituted a significant extra input and accounted for 7 to 29% of the total water input (McJannet et al. 2007). Another recent study on fog deposition, using an eddy covariance system in a Puerto Rican forest, yielded an estimate of ~ 785 mm yr^{-1} (Holwerda et al. 2006). Although the estimation of the additional water input from fog remains technically difficult, these studies show that interception of cloud and fog water may play a very important role in TMF, especially in cloud forests.

Transpiration

The overall annual transpiration rate in forests can be deduced from catchment studies and was 471 mm yr^{-1} in the above-mentioned study in Ecuadorian TMF (Fleischbein et al. 2006). Sap flow studies may provide a more detailed picture of transpiration as they allow for differentiating between species, tree structural characteristics and micrometeorological factors. In the Ecuadorian TMF daily tree sap flow depended mainly on daily radiation measured as photon flux density, and air humidity was important on shorter time scales (Motzer et al. 2005). The data suggest that variation in whole tree water use was more pronounced among trees differing in size and social position (canopy trees vs. understory trees) than among species. In Australia, annual transpiration derived from sap flow measurements in three TMFs ranged from 353 mm yr^{-1} for cloud forest to 591 mm yr^{-1} for montane rainforest with low cloud impact (Mc Jannet et al., in press). Total stand transpiration was controlled by forest characteristics such as stem density, size distribution and sapwood area, and again depended strongly on solar radiation and atmospheric demand for moisture. In general, transpiration in TMFs seems to be quite low as compared with other tropical forests. However, only few studies have been carried out on transpiration by trees in TMFs, and much remains to be done in order to improve our understanding of this subject.

Rainfall interception

As mentioned earlier, re-evaporation of intercepted water is one of the major pathways for water output in TMFs. In comprehensive reviews it was estimated at 20-37% of incident rainfall for TMF not subject to appreciable cloud incidence (Bruijnzeel & Proctor 1995, Bruijnzeel 2001). Even higher values have been published since (Fleischbein et al. 2006, Dietz et al. 2006). It is sometimes suggested that these high values can be attributed to the epiphyte vegetation, which is usually quite abundant in TMFs. However, a modelling study on the contribution of mossy epiphytes to overall rainfall interception in an upper montane forest of Costa Rica suggested that mosses contributed about 6% to total interception (Hölscher et al. 2004). Hence, the hydrological importance of epiphytes in the studied forest was rather limited despite their considerable maximum water storage capacity. This is thought to reflect the fact that under the prevailing rainfall conditions only a fraction of the potential storage is actually available. Low importance of epiphytes for rainfall interception was also found in the Ecuadorian study (Fleischbein et al. 2005). For the latter forest, it was assumed that the high rainfall interception values were attributable to the strong solar insolation characteristic of equatorial sites, the limited impact of fog, low intensity of incident precipitation and additional wind-driven advective energy input (Fleischbein et al. 2006).

Influence of forest use on rainfall interception

Changes in rainfall interception along land use gradients were studied in a TMF region in Central Sulawesi, Indonesia (Dietz et al. 2006). Stands of four management types (natural forest, forest subject to small-diameter timber extraction, forest subject to selective logging of large-diameter timber, and cacao agroforest under trees remaining from the natural forest) were analyzed (12 plots). Rainfall interception was highest in natural forests where 30% (median) of the gross precipitation was re-evaporated back into the atmosphere, and much lower in the three other land use types (18–20%). For individual plots, interception values ranged between -1 to 54% with lowest values in cacao agroforest and highest in natural forest. Interception percentages increased significantly with increasing tree height ($r^2 = 0.63$) (Fig. 2). In a multiple linear regression with tree height and LAI as influencing factors, 81% of the variation in interception percentage was explained. A possible reason for the detected tree height-LAI-throughfall relationship is that under the conditions prevailing in the study region the canopy may not completely dry up between subsequent rainfall events (Dietz et al. 2007). Therefore, the actual water storage at the beginning of a rainfall event would be below its potential. Tall trees increase the vertical distribution of foliage and other canopy components contributing to the canopy water storage, resulting in higher canopy roughness and

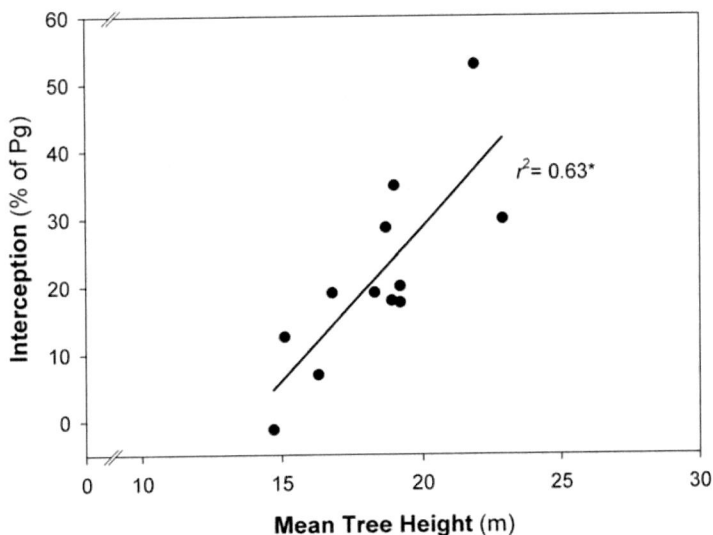

Figure 2. Rainfall interception as a function of mean tree height (dbh $\geq$ 10 cm) in 12 study stands along a gradient of forest use intensity in Central Sulawesi, Indonesia (Dietz et al. 2006). The mean tree height decreased as a result of increasing forest use/disturbance intensity.

a more effective energy exchange with the atmosphere. This, then, would lead to an increased re-evaporation of intercepted water, larger available water storage and, thus, a reduced throughfall in stands with tall trees. It was concluded that forest use reduced rainfall interception as a consequence of reduced canopy height (Dietz et al. 2006). How this affects the ecosystem water budget remains unclear so far. Transpiration seems to a large extend to be controlled by atmospheric parameters and may not significantly increase as a result of forest use. It is thus likely that streamflow increases as a consequence of forest use or disturbance.

In order to predict rainfall interception patterns on a regional scale in this human-dominated forest margin zone in Indonesia, a study based on satellite image analyses was conducted (Nieschulze et al., submitted). Canopy roughness was represented by local maxima of reflectance. The underlying idea was that maximum values indicated tree tops, whereas minima or valleys represented gaps or breaks between tree crowns. For the twelve plots, multiple regression with local maxima as main parameter yielded an r² of 0.84. Application of the model at the landscape level (784 ha) provided realistic results with high interception values being rare and restricted to natural forest stands distant to villages, whereas low interception characterized the intensively used sites close to settlements. (Fig. 3). The 11% regional interception of incident rainfall obtained in this study is quite low and suggests that most of the forest in the study area has been used (Nieschulze et al., submitted).

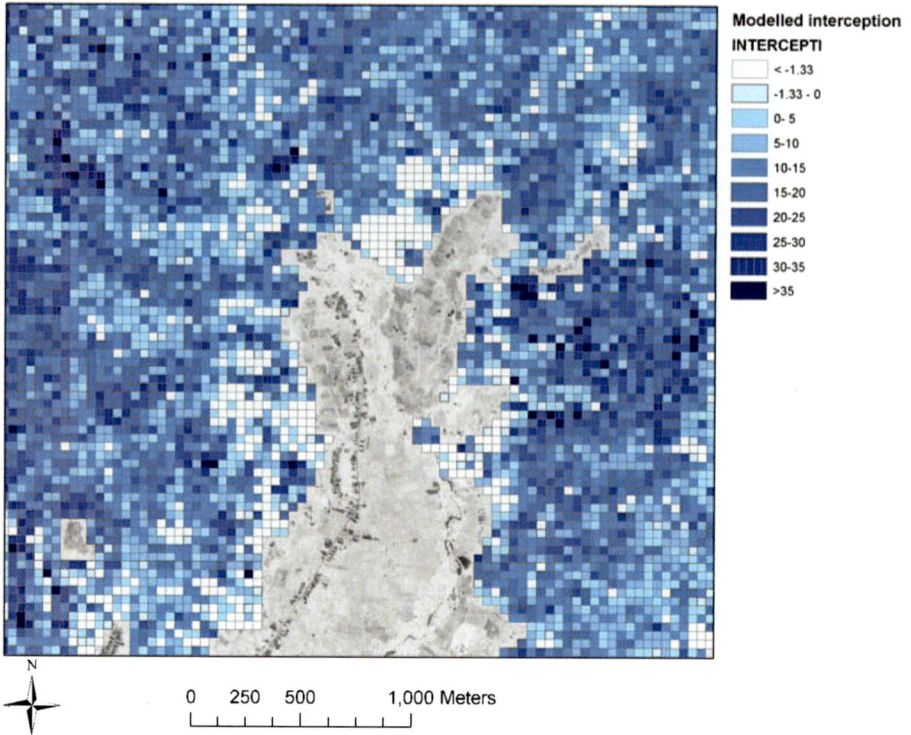

Figure 3. Predicted rainfall interception in a human-dominated forest landscape in Central Sulawesi, Indonesia, based on satellite image analysis (Nieschulze et al., submitted). Settlement areas and paddy rice fields (in the centre and at the bottom) are masked. Other fields include cacao agroforest, logged forest and (some remnant) natural forests.

The study leads to the conclusion that satellite image analysis can successfully be applied for the regional prediction of interception forest areas.

Conclusions

The hydrology of near-natural tropical mountain rainforests is characterized by high rainfall, the possibility of additional water input from cloud interception, low transpiration rates, high streamflow and high rainfall interception rates. Forest use may significantly reduce rainfall interception, which probably leads to even higher streamflow. Remote sensing appears to be a useful tool for predicting rainfall interception at the regional scale.

Acknowledgements. This study was partly conducted in the framework of the joint Indonesian-German research project 'Stability of Tropical Rainforest Margins, Indonesia (STORMA)' funded by the German Research Foundation (SFB 552).

References

Asner GP, Knapp DE, Broadbent EN, Oliveira PJC, Keller M, Silva JN (2005) Selective logging in the Brazilian Amazon. Science 310: 480-482

Bonnell M, Bruijnzeel LA (2005) Forests, Water & People in the Humid Tropics. Cambridge University Press, Cambridge

Bruijnzeel LA (2001) Hydrology of tropical montane cloud forests: A reassessment. Land Use Water Resources Research 1: 1.1-1.18

Bruijnzeel LA, Proctor J (1995) Hydrology and biochemistry of tropical montane cloud forests: what do we really know? In: Hamilton LS, Juvik JO, Scatena FN (eds) Tropical Montane Cloud Forests, Ecological Studies 110. Springer, Berlin, pp 38-78

Dietz J, Hölscher D, Leuschner C, Hendrayanto (2006) Rainfall partitioning in relation to forest structure in differently managed montane forest stands in Central Sulawesi, Indonesia. Forest Ecology and Management 237: 170-178

Dietz J, Leuschner C, Hölscher D, Kreilein H (2007) Vertical patterns and duration of surface wetness in an old-growth tropical montane forest, Indonesia. Flora 202: 111-117

Fleischbein K, Wilcke W, Goller R, Boy J, Valarezo C, Zech W, Knoblich K (2005) Rainfall interception in a lower montane forest in Ecuador: effects of canopy properties. Hydrological Processes 19: 1355-1371

Fleischbein K, Wilcke W, Valarezo C, Zech W, Knoblich K (2006) Water budgets of three small catchments under montane forest in Ecuador: experimental and modelling approach. Hydrological Processes 20: 2491-2507

Hölscher D, Köhler L, van Dijk AIJM, Bruijnzeel LA (2004) The importance of epiphytes to total rainfall interception by a tropical montane rainforest in Costa Rica. Journal of Hydrology 292: 308-322

Holder CD (2003) Fog precipitation in the Sierra de las Minas Biosphere Reserve, Guatemala. Hydrological Processes 17: 2001-2010

Holwerda F, Burkard R, Eugster W, Scatena FN, Meesters AGCA, Bruijnzeel LA (2006) Estimating fog deposition at a Puerto Rican elfin cloud forest site: comparison of the water budget and eddy covariance methods. Hydrological Processes 20: 2669-2692

McJannet D, Wallace J, Fitch P, Disher M, Reddell P (2007) Water balance of tropical rainforest canopies in north Queensland, Australia. Hydrological Processes, DOI: 10.1002/hyp.6618 (Wiley InterScience)

McJannet D, Fitch P, Disher M, Wallace J (in press) Measurements of transpiration in four tropical rainforest types of north Queensland, Australia. Hydrological Processes, DOI: 10.1002/hyp.6576 (Wiley InterScience)

Motzer T, Munz N, Küppers M, Schmitt D, Anhuf D (2005) Stomatal conductance, transpiration and sap flow of tropical montane rain forest trees in the southern Ecuadorian Andes. Tree Physiology 25: 1283-1293

Biodiversity and Ecology Series (2008) 2: 137-162
The Tropical Mountain Forest – Patterns and Processes in a Biodiversity Hotspot
edited by S.R. Gradstein, J. Homeier and D. Gansert
Göttingen Centre for Biodiversity and Ecology

Soil, climate and vegetation of tropical montane forests – a case study from the Yungas, Bolivia

Gerhard Gerold

Landscape Ecology, Institute of Geography, University of Göttingen, Goldschmidtstr. 5, 37077 Göttingen, Germany, ggerold@uni-goettingen.de

Abstract. Floristic composition, structure and functioning of tropical montane rainforests are determined by abiotic and biotic factors. The precise nature of the causal relationships is still a matter of debate. In the framework of an interdisciplinary project aiming at understanding the relationships between vegetation and abiotic factors in the montane forest belts in the northeastern Andes of Bolivia, hypsometric changes in soil and hydrometeorology were studied along an altitudinal transect between 1700–3400 m. The results suggest that the difference in forest stature of the lower and upper montane forest belts is primarily due to the different radiation climate, that of the upper montane and subalpine belts by the pronounced increase in precipitation towards the latter belt. These conditions lead to persistently saturated conditions, high soil acidity and nutrient leaching in the subalpine belt, with Placorthods and Placaquods being the dominant soil types.

Introduction

The Yungas region in the northeastern Andes of Bolivia (north of 18° S) is one of the world's hotspots of biodiversity (Barthlott et al. 1996, Myers et al. 2000). Probably 10,000 vascular plant species occur in the area (of which only ca. 7000 are described), or about 50% of all plant species of Bolivia (Kessler & Beck 2001, Bach 2004). Several plant groups reach their maximum of diversity here, e.g. orchids with about 1170 species and melastoms with ca. 100. According to Beck (1988), the core zone of the Yungas extends from 1200 to 3400 m and includes three elevational vegetation belts, the "lower montane forest" (LMF), the "upper montane cloud forest" (UMCF) and the "subalpine cloud forest" belt (SCF) (see also Bruijnzeel & Hamilton 2000). The elevation of the timber line in the area varies with exposition, relief (i.e. slope), humidity and anthropogenic influence, and reaches maximally up to 3600 m (Beck 1998).

Tropical montane rainforests possess a high relevance for the hydrological cycle and water resources (Bruijnzeel 2005). They provide high water quality and constant discharge for users downstream. Additional water input via fog and precipitation is captured by leaves and epiphytes in the forest canopy. Interception may be about 10–20% of incident rainfall, in exposed areas up to 60–70% (Bruijnzeel & Hamilton 2000), thus reducing the risk of landslides on steep slopes.

With increasing elevation, the typical diurnal climate of the tropics changes by declining air pressure and temperature, reduction of evapotranspiration, and increase of cloudiness, fog and UV-B radiation intensity (Hamilton 1995, Bruijnzeel 2005). These changes are correlated with, a.o., decline of tree height and lack of emergent trees, decreasing leaf sizes, and an increase in scleromorphy and epiphytism. The vegetational changes along the elevational gradient have often been described but the determining factors underlying these changes are still controversially discussed. Explanations for the reduction of tree height with altitude include: 1) the permanently low air saturation deficit in the mountains, reducing nutrient and water uptake (Odum 1970); 2) the reduced irradiation and temperature with altitude, and the increased cloud cover and humidity, impacting transpiration and photosynthetic activity (Grubb 1977, Stadtmüller 1987, Bruijnzeel & Veneklaas 1998); 3) the elevated UV-B radiation in the mountains, causing damage to photosynthesis (Flenley 1996); 4) soil oxygen deficiency due to permanently high water saturation (Hetsch & Hoheisel 1976, Santiago et al. 2000); and 5) high Al-saturation and concentration of phenolic compounds in soil organic matter (Hafkenscheid 2000). Based on work in South Ecuador, Leuschner et al. (2007) found that the altitudinal decrease in tree height is primarily the consequence of a shift in aboveground/belowground carbon allocation patterns: "We hypothesize that the increasing fine root biomass with elevation is caused by an increasing importance of nutrient limitation of tree growth" (p. 227; see also chapter 8 of this volume).

Soil genesis studies in tropical montane forests have identified stagnic and podzolic properties as the principal soil-forming processes (Schawe et al. 2007). Hetsch & Hoheisel (1976) emphasized the importance of hydromorphic processes and classified montane forest soils in the Andes of Venezuela as Spodic Dystropepts. Schrumpf et al. (2001) reported increasingly aquic conditions and placic horizons with increasing elevation in the Andes of southern Ecuador. Towards high elevations (ca. 3050 m), however, shallow and less developed soils prevailed. Generally, our knowledge of soil genesis in tropical montane forests is still fragmentary, information on soil-forming substrates is sometimes lacking and comprehensive soil surveys have not been conducted (Roman & Scatena in press).

The present study focuses on the relationships between climate, soil and vegetation in the Yungas of Bolivia. The principal altitudinal belts in the area have been described but an integrated analysis of climate, soil and vegetation changes along the elevational gradient had not been carried out. Research questions included:

1)   How do soil and climate parameters change hypsometrically?
2)   How do nutrient dynamics, as measured by precipitation and litterfall, change along the gradient?

3)   Which interrelations can be detected between vegetation units and abiotic
     factors?

The results will be discussed in the light of the question as to which parameters are
responsible for the decrease of tree height with elevation.

Figure 1. View of the Yungas and Cotapata National Park (background) from Coroico,
with the forest line at ca. 3400 m (from Kellner 2005).

Material and Methods

Research area. The present study was carried out along a transect on a south-easterly orientated slope in Cotapata National Park in the northeastern part of the Bolivian Andes, ca. 80 km north of La Paz (16°09'S, 68°55'W; Fig. 1). The eastern Andes are part of the seasonal humid tropical zone, having a maximum of precipitation during summer and short drier periods in winter (July, August). The transect extended along a ridge from 1700 m to the timberline at 3400 m and encompassed the three montane forest types mentioned above that are characteristic of the Yungas region (Fig. 2). The longitudinal slope profile was mostly linear and had an average inclination of 25° to 30°, in some areas exceededing 40°. The geological substrate consisted of Ordovician meta-siltstones, slates and meta-sandstones (Mapa geologico de Coroico, Hoja 6045). Soils were characterized by high acidity, low cation exchange capacity and high aluminum-saturation in the mineral hori-

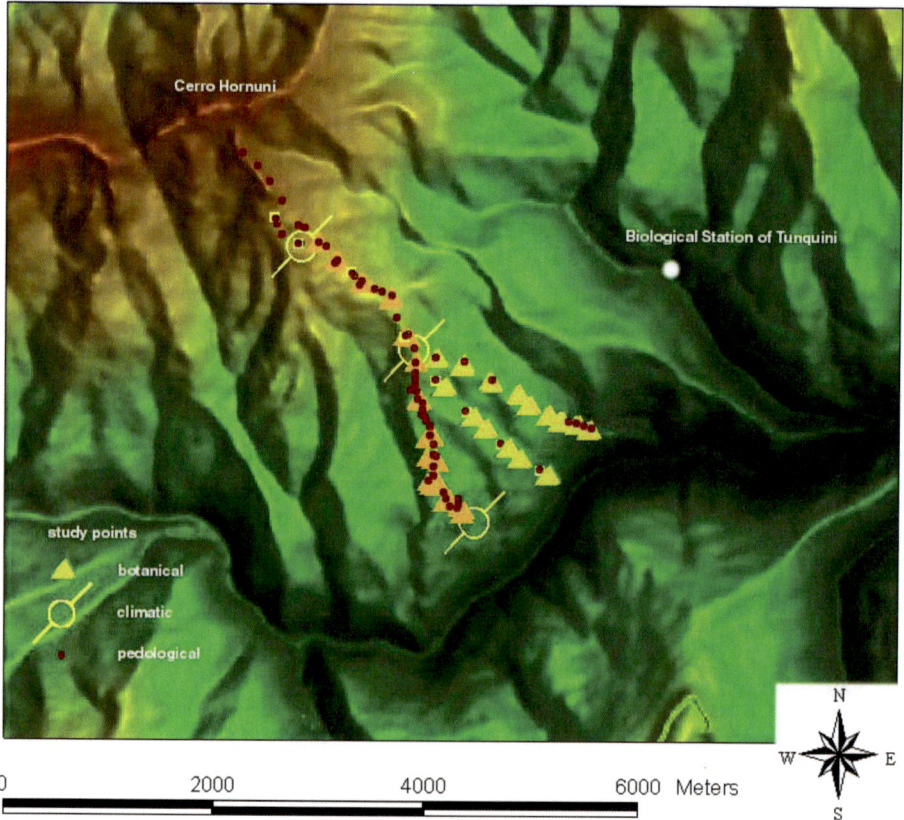

Figure 2. Research area in Cotapata National Park, Bolivia, showing the altitudinal transect studied. Dots: soil sampling sites. Triangles: vegetation sampling sites. Circles: climate stations.

zons, and organic layers of 15 to 45 cm (LMF: 15–35 cm; UMCF: 25–40 cm; SCF: 20–35 cm; see Schawe 2005, 2007).

All three elevational vegetation formations recorded from the park by Beck (1988) – lower montane forest (LMF), upper montane cloud forest (UMCF), subalpine cloud forest (SCF) – occurred along the transect (Bach 2004; Table 1). The limits of the three belts were not sharp, however (Bach 2004). LMF extended up to ca. 2150 m and had a dense, ca. 20 m high canopy, and emergent trees up to 35 m. *Myrsine coriacea* (Sw.) R.Br. ex Roem. & Schult. was the dominant tree while the fern *Blechnum ensiforme* Liebm. and the aroid *Philodendron ornatum* Schott dominated in the understorey. Above 2150 m LMF changed to UMCF. The latter forest type was subject to persistent cloud incidence and was dominated by the tree species *Podocarpus oleifolius* D.Don ex Lamb., *Weinmannia crassiflora* Ruíz & Pavón and *Clusia multiflora* Kunth. The canopy of UMCF was lower than that of LMF and typically reached up to 15 m. Abundance of epiphytes, mostly bryophytes (mosses, liverworts), was noteworthy; terrestrial bryophyte cover was also relatively high and estimated at about 45%. Above 2800 m UMCF gave way to SCF. The dominant tree species in SCF, *Myrsine coriacea* and *Podocarpus rusbyi* J.Buchholz & N.E.Gray, reached heights of ca. 10 m and did not form a closed canopy. Terrestrial bryophyte cover increased to 75% and was dominated by *Sphagnum* species. Scleromorphic shrubs such as *Miconia*, *Gaultheria*, *Gynoxys*, *Ilex* and *Escallonia myrtilloides* L.f. were abundant. In melastoms the proportion of sceromorphic species increased from 11% in UMCF to over 50% in SCF (Table 1). In all analyzed plant groups (ferns, melastoms, aroids, bromeliads, palms, cacti) α-diversity decreased significantly with elevation. Highest overall species diversity occurred at 1900 m (Bach 2004, Bach et al. 2007).

Table 1. Vegetation parameters along the transect (from Bach 2004).

vegetation parameter	1800	2000	2200	2400	2600	2800	3000	3200
height of emergent trees [m]	28	31	25	24	-	-	-	-
mean canopy height [m]	20	20	17	18	15	11	12	6
epiphyte cover [%]	15	30	45	55	65	45	40	35
scleromorphic species in melastoms [%]	15	30	30	35	33	60	75	100

Climate. Meteorologic measurements were carried out over a 2,5 year period, with continuous readings from October 2001 until October 2002, in forest clearings at 1850 m (LMF), 2600 m (UMCF) and 3050 m (SCF) (Fig. 2). Forest clearings were 2500 m² (LMF) and 900 m² (UMCF, SCF) in accordance with park rules and relief situation, causing reduction of wind speed and horizon limitation at sunrise and sunset. Rainfall (mm) was measured at 1 m above ground level using a tipping-bucket recording gauge (ARG 100, EM) with 0.2 mm resolution per tip. Short-wave radiation was measured with a pyranometer (8101, Schenk), net radiation with a net radiometer (NR-lite, Kipp), and photosynthetic photon flux density with a quantum sensor (DK-PHAR, Deka). All radiation sensors were placed at 2 m height on supporting arms at 1.5 m from the mast to avoid shading of the instruments. Air temperature and relative humidity were measured with sensors placed within a Gill-type radiation shield at 2 m height and, additionally, at 0.5 m, 2 m and 15 m height in the forest at 200 m elevational intervals from 24th November until 15th April 2001. Measured relative humidity values in excess of 100% were set to 100%. Wind speed (A100 anemometer, Vector Instruments) was also recorded at 2 m height, i.e. well below the level of the surrounding canopy. Instruments were sampled at 10 min intervals except for wind speed which sampled every minute. Interception loss in the three vegetation belts was measured in 2005 during a three-month period (April–June) coinciding the transition from rainy to dry season (10 roved gauges, weekly on 500 m² plots).

Data were processed by a Delta T data-logger system; averages and standard deviations were calculated over 30 min periods. Reference evaporation (potential evapotranspiration) was calculated from radiation, temperature, humidity and wind speed data using the modified Penman equation (Doorenbos & Pruitt 1988). The degree of reduction in Rs due to the presence of fog and clouds was estimated from a comparison with potential radiation values as calculated using a topography- and exposition-based radiation model (Böhner et al. 1997). Direct radiation was calculated by the sun inclination and the angle between sun inclination and topographical area. The amount of diffuse radiation was calculated by sun altitude and direct sun radiation, including air pressure.

Soil. Soils were studied along the main transect (1700–3400 m) and along two parallel short transects from 1800 m to 2600 m, all located on extended ridges (Fig. 2). Excavated soil pits were sampled at 100 m intervals along the long transect and at 200 m intervals along the two short ones until reaching 1 m depth or lithic contact. The sampling scheme resulted in 26 profiles. Soil profiles representative of elevational soil zones were selected based on soil texture, colour and pH. Soils were classified following the "Soil Survey Staff" (2003). Soil samples were passed through a 2 mm mesh. For texture analysis, samples were prepared by destroying organic matter with 10% H_2O_2 and dispersion by $Na_2P_4O_7$. Grain size distribution of the fine earth (< 2 mm) was determined by sieve (≥20 μm) and pipette (< 20 μm). Bulk densities of the mineral horizons were measured by weighing undisturbed soil cores (100 cm^3) after drying at 105 °C. Bulk densities of the organic layer were determined by extracting samples of the organic layer of 20×20×10 cm and drying at 105 °C. Soil pH was analyzed potentiometrically (GPHR 1400) in 0.01 M CaCl$_2$ with a soil–water ratio of 1 : 2.5 for mineral samples and a soil–water ratio of 1 : 5 for organic samples. Total C and N were determined by a Leco CHN-analyzer. Effective cation exchange capacity was determined by extraction with NH$_4$Cl (Luer & Böhmer 2000) and measured by inductively coupled plasma optical emission spectrometry (ICP-OES, Optima 4300 DV, Perkin Elmer). Base saturation percentage was expressed as the sum of the cations divided by ECEC*100. Total Fe concentration was determined by pressure digestion with HF, HNO$_3$ and HCl (Heinrichs et al. 1986). Oxalate-extractable Fe and Al were analyzed according to Schwertmann and dithionite-extractable Fe and Al according to Mehra & Jackson (Schlichting et al. 1995). Iron and Al concentrations of all extracts were measured by ICP-OES.

Water content of the ectoorganic soil layer was measured daily over a period of three weeks during the dry season (September–October 2002) at 200 m elevation intervals. Three samples were taken at 10 cm depth, weighed, dried at 105 °C and reweighed. Gravimetric water contents were expressed in grams of water per gram of oven-dried soil.

Nutrient input. To analyse nutrient fluxes, litter and bulk rainfall deposition samplers was installed at the climatic stations. Seven litter samplers were randomly placed in an area of 1 ha during April 2003 – April 2004 in each vegetation belt. Every four weeks litter was collected, air-dried, separated into leaves, branches and epiphytes, and dry weight estimated. Weight-representative samples were grinded and total element concentration determined by pressure digestion with HF, HNO$_3$ and HC$_l$ (Heinrichs et al. 1986), CHN-analysator (C$_t$, N$_t$) and ICP-OES (P, S, Ca, Mg, K, Na, Mn, Al, Fe, Zn). Nutrient concentrations in rainfall were measured during April–July 2005 (Kellner 2006) by weekly collection of incident rainfall (3 rain gauges each plot) and throughfall (10 roved bulk sampler with monthly change of location; Veneklaas et al. 1990). After measuring pH and EC, rain samples were filtered using 0,45 μm fibreglass and stored at 4 °C. Element were analysed by ICP-OES (Na, K, Mg, Ca) and IC-Methrom (Cl, NO$_3$, PO$_4$, SO$_4$). Concentrations of Mn, Fe, Pb, Mo and Cu were below ICP-OES detection limit.

Results

Hydrometeorological characteristics. Monthly incoming radiation was high and decreased with increasing elevation, UMCF at 2600 m receiving about 33% less short-wave radiation than LMF (1850 m) and SCF (3050 m) ca. 5% less than UMCF (Table 2). Reduction of short-wave radiation due to fog and clouds was 37% in LMF, 58% in UMCF and 62% in SCF (Gerold et al. 2007). Thus, radiation reduction was most pronounced in SCF where it exceeded 50% on 306 days/y and reached as much as 90% during a 4-day period. Net radiation showed similar patterns, with minor seasonal deviations. Mean daily net radiation (Rn, 6–18 h) was 63% of Rs in LMS and reduced to 50% of Rs in SCF due to decreasing incoming radiation und increasing outgoing radiation during nights. During clear days maxima between 18 MJ (SCF) and 26 MJ m^{-2}*d (LMF) were reached, whereas for enduring cloudy days 2 MJ (SCF) and 3 MJ m^{-2}*d (LMF) are characteristic (Schawe 2005).

Table 2. Radiation and temperature in the three montane forest types studied (October 2001 – October 2002).

climate parameter	LMF 1850 m	UMCF 2600 m	SCF 3050 m
mean daily short wave input Rs [MJ m^{-2}]	15.5±5.4	10.4±4.3	9.9±3.8
Rs range [MJ m^{-2}]	4.6-29.8	3.3-28.1	3.1-24.8
average daily photosynthetic photon flux density PPFD [µmol m^{-2} s^{-1}]	749	471	455
average daily maximum PPFD [µmol m^{-2} s^{-1}]	1,500	910	790
average daily maximum PPFD on full cloudy days [µmol m^{-2} s^{-1}]	505	410	405
average daily temperature [°C]	16.8±1.7	15.6±1.6	10.0±1.5
daily amplitude of temperature [°C]	9.8	6.6	5.9
average minimum and range of temperature [°C]	13.3 (7.7-17.2)	10.2 (5.7-14.0)	7.4 (2.1-10.8)
average maximum and range of temperature [°C]	22.8 (14.8-29.0)	16.8 (8.8-23.8)	13.3 (6.9-18.5)

Air temperature and photosynthetic photon flux density decreased with increasing elevation along with reduction of net radiation and increase of cloudiness; small yearly amplitudes in temperature were sustained (Table 2). Overall temperature decline between LMF and SCF (0.59 °C per 100 m) was close to the wet adiabatic lapse rate (0.56 °C per 100 m; Lauer 1975) and recorded lapse rates between vegetation belts were similar (0.53 °C per 100 m between LMF and UMCF; 0.62 °C per 100 m within the cloud belt; Table 2). The diurnal range in temperature decreased with increasing altitude. Average relative humidity increased slightly from 90% in the LMF to 96% in the UMCF and 97% in the SCF whereas the average diurnal range decreased with altitude, from 30% at 1850 m to 5% at 3050 m. The measurements in the forest showed a distinct increase of daily average minimum relative humidity at 2 m height, from 87% in the LMF to 95% in the UMCF (Pareja 2005). The closed forest canopy and high degree of cloudiness causes maximum humidity all year round between 2200–2600 m.

Rainfall measurements showed the occurrence of a relatively dry season during May–September and a main rainy season during October–April. Annual rainfall was high and increased considerably with altitude (Table 3). UMCF received 1.7 times more rainfall than LMF, and SCF 2.2 times. The number of days with < 1 mm/day of rainfall decreased from 72% in LMF to 60% in SCF but those with 2–10 mm/day increased from 15% to 23%. The length of continuously dry periods decreased from 9 days at 1850 m to 7 days at 2600 m and 5 days at 3050 m.

Daily reference evaporation (PET) was very similar in UMCF and SCF (1.1–1.3 mm), and almost three times lower than in LMF; variation in daily values was substantial at all three sites (Table 3). Average hourly PET was 0.14 mm in LMF but only 0.05 mm in UMCF and SCF. Rainfall exceeded PET 1.9 times in LMF, 8.6 times in UMCF and 13.2 times in SCF (Table 3).

Table 3. Rainfall and evapotranspiration in the three montane forest types studied (October 2001 – October 2002).

climate parameters	LMF 1850 m	UMCF 2600 m	SCF 3050 m
rainfall [mm yr^{-1}]	2310	3970	5150
interception E_i [mm yr^{-1}]	494	1040	1308
annual evapotranspiration PET [mm yr^{-1}]	1190	462	403
daily PET [mm yr^{-1}]	3,3	1,3	1,1
range daily PET [mm yr^{-1}]	0,3-8,6	0,07-6,4	0,04-5,7

Soils. In LMF (1700–2200 m) silty, brown soils with shallow ectoorganic horizons (10–30 cm) dominated. Humic Dystrudept was the representative soil type (Fig. 3). The weakly developed Inceptisols gave way to more strongly altered soils in UMCF (2200–2700 m). Soils in this belt usually differed from those in LMF by the thicker ectoorganic horizon (30–40 cm) and the presence of a grey E horizon and a

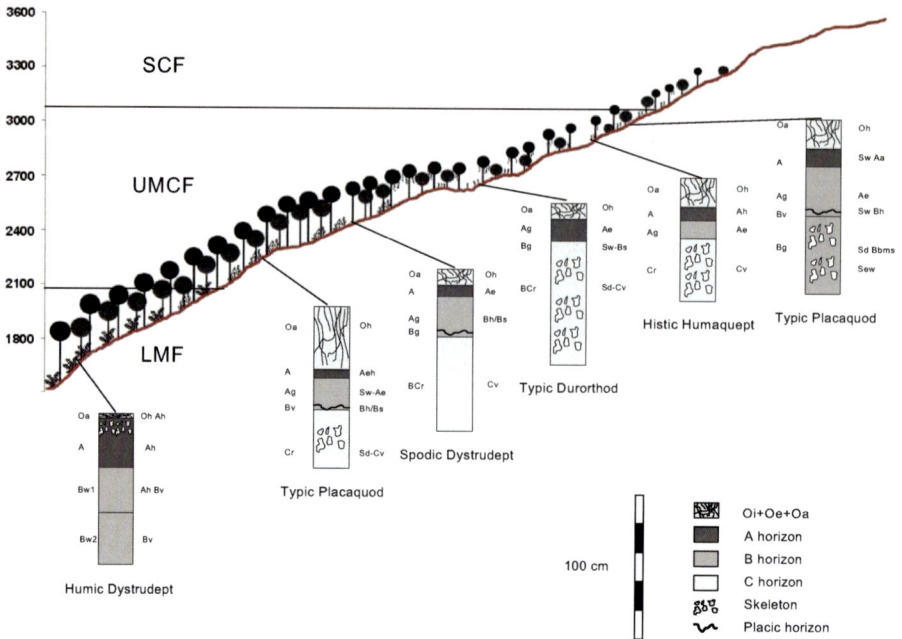

Figure 3. Representative soil types (USDA) along the transect.

Bhs one (placic horizon). Typic Placorthod was the representative soil type in this belt (Fig. 3). Soils with hydromorphic properties (mottles, Fe concretions) became increasingly more prominent above 2500 m and dominated in SCF above 3000 m. Typic Placaquod was the representative soil type in the latter belt (Fig. 3); the thick (35 cm) ectoorganic horizon was followed by a dark grey silty to loamy Ag one. A placic horizon limits rooting depth; below this horizon, aeration was improved.

Moisture content in the organic layer increased (although somewhat irregularly) with elevation from 2.5 g g^{-1} at 1800 m to 6.3 g g^{-1} at 3000 m (Fig. 4). The main increase was from 1800 to 2100 m (0.5 g g^{-1} per 100 m elevation interval), coinciding the transition from LMF to UMCF, and from 2200 to 2700 m (0.5 g g^{-1} per 100 m). Mineral soils also showed significant differences in moisture content between LMF and the other two vegetation formations. In the upper soil horizon (Ah) in LMF, soil moisture remained at field capacity (48%) during the rainy season (February–June) and varied between 30–45% in other months depending on daily rainfall events (Fig. 5). Seasonal distribution of moisture content in the upper soil horizon in both UMCF and SCF was relatively constant. In SCF soils were saturated throughout the year, in UMCF only during the rainy season, remaining at field capacity during the rest of the year (Fig. 5).

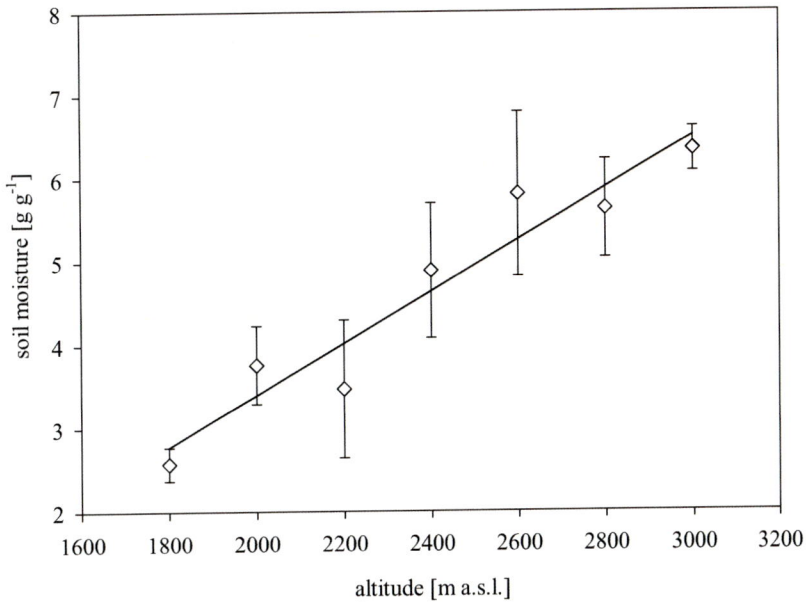

Figure 4. Soil water in the organic layer along the transect.

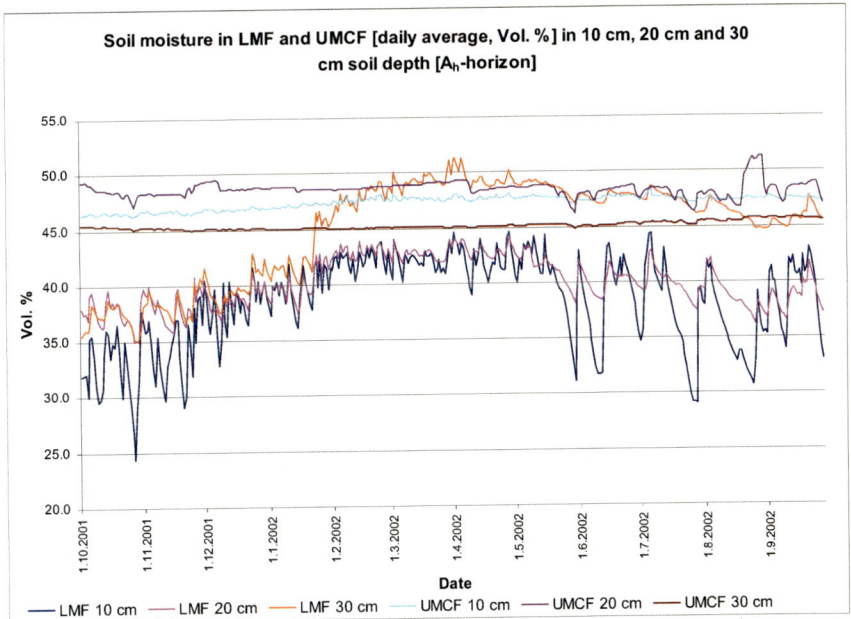

Figure 5. Soil moisture in LMF and UMCF in 2002 (field capacity at 48%; permanent wilking point at 18%).

Soils were generally acidic and pH in organic layers ranged from 2.4 to 3.5. In the mineral horizons, pH increased gradually with depth (Schawe et al. 2007). Between 1850–2000 m the pH of the A horizon was around 4 and between 2000–3001 m around 3; above 3100 m it increased again to 3.5. The elevational course of the pH of the mineral subsoil differed little from that in the organic horizons. The effective cation exchange capacity (ECEC) in the research area was very low, being < 15 $cmol_c$ kg^{-1} in the A horizons and < 25 $cmol_c$ kg^{-1} in the ectoorganic horizon. Between 1800–2100 m average ECEC was 9.3 $cmol_c$ kg^{-1}, decreasing to 7.5 $cmol_c$ kg^{-1} above 2100 m and a mere 6.3 $cmol_c$ kg^{-1} in SCF. ECEC in the A horizon of LMF and UMCF was dominated by Al, occupying more than 80% of total exchange capacity (Schawe et al. 2007). High concentrations of Al or low base saturation have also been noted in tropical montane rainforests by Hafkenscheid (2000), Fölster & Fassbender (1978), Grieve et al. (1990).

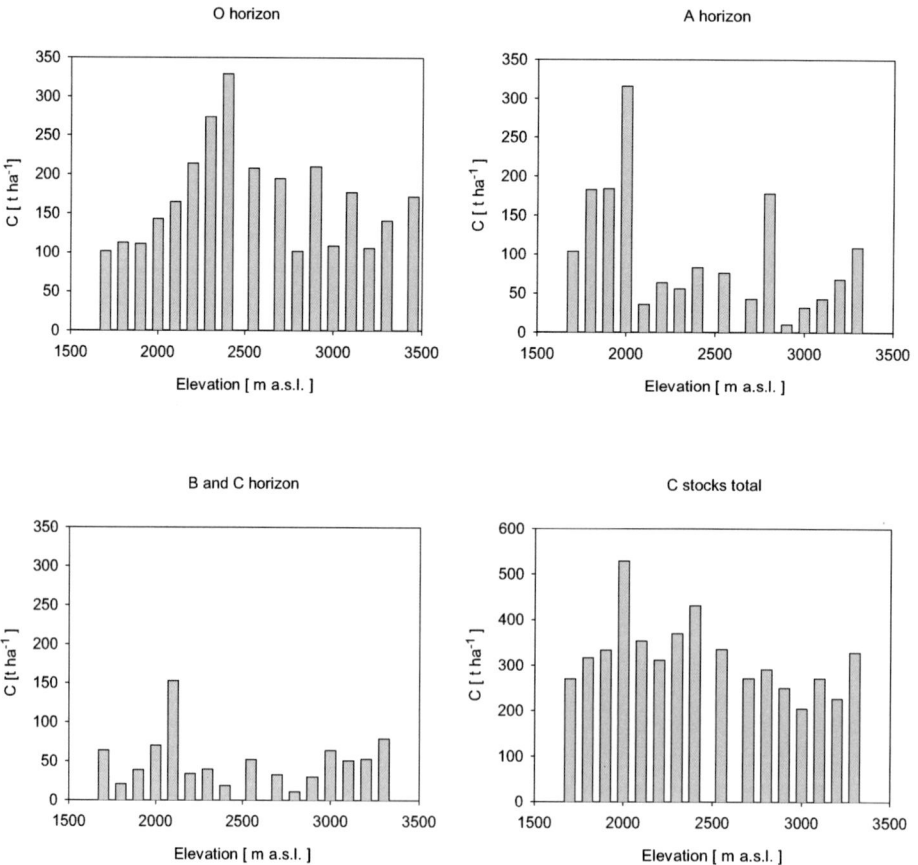

Figure 6. Carbon stocks (t ha^{-1}) in different soil horizons along the transect.

In LMF < 30% of the soil organic carbon (SOC) was accumulated in the ectoorganic horizon, at higher elevations more than 50% (Fig. 6). There is, however,
no continuous rise of total carbon stocks with altitude, rising until 2400 m in the O
horizon and until 2000 m in the mineral soil. Above these elevations, no elevational pattern could be detected. Total SOC stocks ranged from 220 to 530 t ha^{-1}
with average values of 363 t ha^{-1}. Maximum values (over 350 t) were measured
between 2000–2400 m due to thick ectoorganic layer of 35–40 cm at these elevations (Schawe 2005).

Nutrient input. Total litter input decreased from LMF (12.2 t ha^{-1} a^{-1}) to UMCF
(5.3 t) by 57% and by a further 12% to SCF (4.7 t). The difference between LMF
and the other two vegetation belts was significant ($p < 0.05$). The leaf litter part
was 69–71% in all three belts but seasonal variation differed (Table 4). Leaf and
wood litter reached maximum amounts at the end of the dry season (Septem-
ber/October); in UMCF and SCF the amount of leaf litter varied at ca. 0.3
t/ha*month. The vegetation belt with the highest abundance of epiphytes produced highest epiphyte litter rates (0.26 t/ha*a, Table 4) with a clear maximum
from July to November. Fruit litter input showed a clear seasonal signal in SCF
with maxima in July and August (45–60 kg ha^{-1} month^{-1}), and had the highest rate
with 0.25 t ha^{-1} a^{-1}in SCF (5% of total litter input). Compared to other Andean
mountain forests, litter input of LMF is at the upper limit, for UMCF and SCF
total amount is within the standard range (4.3 – 7.4 t ha^{-1} a^{-1}) (Hafkenscheid
2000).

Table 4. Litter input (t ha^{-1} a^{-1}) with epiphytes ( ), stock of Oi-horizon (t ha^{-1}) and mineralization rate (k_{oi}) in tropical montane forests.

country	forest type	altitude [m]	litter input [t ha^{-1} a^{-1}]	Oi-stock [t ha^{-1}]	k_{oi}
Bolivia (this study)	LMF	1900	12.2 (0.12)	5.7	2.1
Bolivia	UMCF	2600	5.3 (0.26)	6.0	0.9
Bolivia	SCF	3050	4.7 (0.03)	6.3	0.7
Venezuela[1]	LMF	2300	7.0	38.0	0.2
Ecuador[2]	LMF	2100	8.5-9.7	7.7-8.8	1.1
World[3]	UMCF	1550-3370	4.3-7.4	6.3-38.0	0.18-1.04

[1] Steinhardt 1976, [2] Yasin 2001, [3] Hafkenscheid 2000

Table 5. Nutrient input with annual litterfall in tropical montane forests (kg ha^{-1} yr^{-1}).

country	forest type	altitude [m]	N	P	Ca	K	Mg	Al	S
Bolivia (this study)	LMF	1900	218	11.0	123	48.4	46.6	15.5	18.0
Bolivia	UMCF	2600	57	3.4	36	14.6	12.5	6.9	4.6
Bolivia	SCF	3050	46	2.7	39	14.5	11.7	4.3	3.9
Venezuela[1]	LMF	2300	69	4.0	43	33.1	14.2	9.3	-
Ecuador[2]	LMF	2100	165-201	8.8-14	113-154	57-77	34-52	4.7-12.0	20-23
world[3]	LMF	610-2600	23.9-100.5	0.7-7.73	16.3-119.2	4.6-59.3	7.6-25.0		
world[3]	UMCF	1550-3370	28.9-80.8	1.1-2.9	6.7-50.2	5.7-33.0	6.3-18.2		

Table 6. Nutrient input with incident rainfall and throughfall in the three montane forest types studied (kg ha^{-1} a^{-1}) (after Kellner 2006).

	LMF incident rainfall	LMF throughfall	UMCF incident rainfall	UMCF throughfall	SCF incident rainfall	SCF throughfall
TOC	77.5	127.3	80.5	271.7	2.3	148.0
TNb	5.7	10.9	4.3	7.5	5.0	4.7
NO$_3$-N	8.1	6.2	7.2	1.9	7.5	3.1
SO$_4$-S	68.5	64.8	150.8	115.5	196.9	153.0
Na	46.0	48.8	98.0	87.4	119.3	102.0
Cl	23.2	25.1	39.8	39.8	55.7	43.9
Ca	7.1	15.3	11.6	11.9	11.9	12.5
Mg	0.4	4.6	1.4	2.4	0.9	1.9
K	1.0	18.6	5.6	5.2	2.6	3.2
PO$_4$-P	1.1	20.1	0.7	0.7	1.2	1.1

Litter nutrient concentrations (mg/g) generally decreased from LMF to UMCF and SCF with significant differences (p < 0.05) for N, P, Ca, K, Mg, S and Mn (Table 8 in Schawe 2005). We found a highest decrease for N (45%!), K, Mg and S from LMF to the other two belts. Nutrient input was 3–4 times higher in LMF than in UMCF and SCF due to higher litter input (Table 5). Furthermore, large differences between LMF and SCF (70–90%) exists for N, P, Ca, Mg, K, S. Comparing the mineralization rate of the litter with the element specific rate, we found for K a double to triple, for Mg a double and for Ca, P and S a 1.5 time higher mineralization rate in all vegetation belts.

Concentrations of most nutrients in incident rainfall were low and without significant differences between the altitudes; variation in relation to single rainfall events and season was high. Significant correlations ($p < 0.05$) between increase in rainfall and decrease in nutrient concentration were only found for TN_b and NO_3. The increase of sulfate with elevation (Table 6) was presumably caused by air masses from the Peruvian-Bolivian altiplano, holding high concentrations of nitrate and sulfate (Kellner 2006).

Discussion

Radiation. Average R_s values from LMF in the study area are comparable to those recorded from southern Ecuador (Motzer 2003) and from Jamaica (Hafkenscheid 2000) under conditions of light cloud incidence. Under more cloudy conditions at 900–1050 m in Puerto Rico, Holwerda (2005) obtained average daily R_s values of 9.7 ± 4.0 MJ (stunted ridge-top cloud forest) and 10.5 ± 4.8 MJ (upper montane palm forest). The latter values are very similar to those of UMCF and SCF in the study area. Holwerda (2005) estimated that R_s in the Puerto Rican cloud belt was reduced by about 50% compared to potential radiation input, which, again, is similar to the calculated 58–62% reduction in the study area, thereby confirming the strong cloudiness of the Bolivian UMCF and SCF sites.

Changes in photosynthetic photon flux density (PPFD) along the elevational gradient reflected the measured changes in R_s (Table 1). Average daily PPFD totals in New Guinea varied from 38 mol m^{-2} at 1100 m to 22 mol at 3480 m (Körner et al. 1983) and were similar to values observed in the present study area. A 500 µmol threshold of light saturation in tree crowns (*Cyrilla racemiflora, Clethra occidentalis*) in a Jamaican UMCF was proposed by Aylett (1985). PPFD values in LMF exceeded this value for 64% of the time vs. 52% in UMCF and SCF. Applying the lower limit of 200 µmol (Garcia-Nunez et al. 1995) would increase these percentages to 76% in LMF, 72% in UMCF and 76% in SCF. Even under fully overcast conditions, PPFD remained above 200 µmol during 8 h in LMF, 6.5 h in UMCF and 7 h in SCF. According to Körner (1982) short periods of intense PPFD are more important for the growth of alpine sedges than low average PPFD levels over longer periods. The average daily maximum PPFD values in the study area were all well above the threshold values discussed above, also during fully cloudy skies (Table 1). The data suggest that despite the marked difference in radiation levels between LMF and the other two vegetation units, reduction in PPFD is unlikely to be a prime cause for the reduction of forest stature with altitude.

Interception and evaporation. Measurements with recording troughs (Juvik & Nullet 1995) or roving gauges (Hafkenscheid 2000) have shown interception losses (E_i) of 25% in SCF (Hawaii) and 22% in UMCF (Jamaica). Much higher E_i value have been recorded from Venezuelan cloud forest at 2300 m (45%; Ataroff 1998) and Ecuadorian cloud forest (39%; Yasin 2001). Rainfall partitioning in relation to forest use or disturbance by logging was measured in Sulawesi (Indonesia) by Dietz et al. (2006). Rainfall interception was highest in the natural forest (median 30%) and much lower in three other forest use types (median 18-20%). In a recent review of interception studies in lower and upper montane cloud forests, however, Bruijnzeel (2005) reported E_i values ranging from 8 to 29%. The median interception values measured in the study area (21% in LMF, 26% in UMCF, 25% in SCF; Table 1) match the 25–29% average given by Bruijnzeel (2005) for tropical montane cloud forests (see also chapter 9 of this volume).

In contrast, our annual PET values from UMCF (462 mm year[-1]) and SCF (403 mm year[-1]) cannot be compared with the water-budget based ET values from similar forests elsewhere (310–390 mm year[-1]; Bruijnzeel a. Proctor 1995) because only apparent ET values can be obtained using the latter approach due to the confounding effect of unmeasured cloud water inputs. Evapotranspiration (ET) estimated by the annual difference of throughfall + stemflow and streamflow in the LMF in Ecuador range between 315–369 mm (Yasin 2001). Estimation of E_t (transpiration, 980 mm year[-1]) of a tall LMF in Venezuela by energy budget calculations leads to ET 2.310 mm year[-1] because of high interception evaporation, which seems rather high compared with total evapotranspiration values provided by Bruijnzeel (2005; 1050–1260 mm). Based on direct measurements of water budget components, Holwerda (2005) estimated annual ET of 674 and ca. 480 mm for the cloud-affected forests in Puerto Rico cited earlier. ET values calculated based on the soil water budget at the hydrometeorological stations (grass and herb soil cover) in the study area, were 406 mm in LMF and 38 mm in UMCF. These unrealistic measures are illustrative of the problems with these measurements in the cloud forest, and of the major influence of the ectoorganic layer on the hydrology of the forest. Accurate measurement of fog and cloud water input is difficult (Bruijnzeel 2005; see also chapter 9 of this volume). Our throughfall measurements in 2005 with roving gauges indicate a cloud water input (throughfall > incident rainfall, net precipitation method) of at least 7% (0.17 mm/d) in LMF, 13% (0.54 mm/d) in UMCF and 15% (0.77 mm/d) in SCF. Using the same method, Hafkenscheid (2000) reported 0.53 mm/d for UMCF of Jamaica, and Holder (2003) 0.5–1.0 mm/d for cloud forest of Guatemala.

Transpiration rates in the two Puerto Rican cloud forests studied by Holwerda (2005) were 1.33 ± 0.95 mm day[-1] and 0.81 ± 0.97 mm day[-1], respectively, and were comparable to the daily PET values from UMCF resp. SCF in the study area (Table 2). Although these rates are low, even lower rates (< 0.5 mm day[-1]) of xylem water movement have been demonstrated in a Hawaiian cloud forest (Santiago et al. 2000). The idea of reduction of montane forest stature and productivity by reduced transpiration rates (Odum 1970, Kitayama & Mueller-Dombois 1994) has been refuted on theoretical grounds by Grubb (1977) and Tanner & Beevers (1990). Recent evidence suggests that the decline in above-ground forest stature with elevation reflects increased investment in below-ground productivity, possibly in response to gradually more adverse edaphic conditions (Leuschner et al. 2007; see also below).

Precipitation and soil moisture. Precipitation (P) in the study area increased with a factor 1.7 in UMCF and 2.1 in SCF as compared with LCF. Increase in P above 2000 m is not typical for all tropical regions. According to Lauscher (1976), P is often highest between 1000 m and 2000 m in the tropics. Vis (1986) and Veneklaas & Ek (1990) reported a decrease of annual P from 3150 mm at 1700 m to 1700 mm at 3050 m in the central Cordillera of the Colombian Andes. On Mount Kinabalu (East Malaysia) changes in P with elevation seemed less pronounced (Kitayama 1992). However, the observed increase in P with elevation in the study area in the eastern Andes of Bolivia is matched by other observations in the eastern Andes. Weischet (1969) reported increased P between 2600 and 3300 m in the eastern Cordillera of the Colombian Andes, due apparently to the occurrence of a secondary maximum of precipitation caused by heating of the high mountain plateau and forced convective precipitation on the upper slopes. Also, a marked increase in P from ca. 2000 mm at 1900 m to ca. 5000 mm at 2380 m and 4500 mm at 3060 m has been recorded in the Andes of southern Ecuador (Leuschner et al. 2007). Apparently, local metereological conditions play an important role here.

The pronounced changes in the climatic water balance (i.e. P minus PET) resulted in major changes in soil moisture content along the altitudinal gradient. In LMF waterlogging of soils was absent in the Ah-horizon (10–20 cm depth) but reached field capacity during the main rainy season at 30 cm depth (Fig. 5). In UMCF field capacity was exceeded all year through and in SCF soils were permanently saturated. Similar patterns of precipitation excess and soil wetness with elevation have been reported for northern Colombia (Herrmann 1971), Mount Kinabalu (Kitayama 1992) and Puerto Rico (Silver et al. 1999, Holwerda 2005).

Persistently wet conditions have been reported for some, but certainly not all tropical cloud forest soils (Bruijnzeel & Proctor 1995, Roman & Scatena in press), providing further illustration that not all cloud forests show equal hydrological behaviour (Bruijnzeel 2005). However, increases in the degree and duration of soil waterlogging are often paralleled by a decrease in montane forest stature (Hermann 1971, Hetsch & Hoheisel 1976, Bruijnzeel et al. 1993, Silver et al. 1999, Santiago et al. 2000). Interestingly, such parallels have been observed both at high and low elevations, suggesting that precipitation excess overrides temperature effects. Where waterlogging is particularly persistent, as in the investigated SCF, aerial roots tend to be particularly common (Bach 2004) as also observed in other very wet cloud forests (e.g. Gill 1969, Lyford 1969).

With the FDR soil moisture measurements, calculation of water uptake from the A-horizon with the soil water budget method was possible (ET). Without water uptake from the ectoorganic horizon (the principal root zone in UMCF and SCF), ET in LMF was 406 mm in 2002. Taking into account interception loss (Ei) total ET was 900 mm/year, which is comparable with post-1993 water budget studies in Puerto Rico and Jamaica (Bruijnzeel 2005).

Soil development. Dominating soil-forming processes change from LMF to UMCF and SCF with podzolization (grey E horizon and Bhs horizon), and hydromorphic properties become increasingly important towards higher elevation. These observations correspond with those of Schrumpf et al. (2001) who reported increased water saturation as well as the occurrence of placic horizons of soils in the Ecuadorian Andes. Whitmore & Burnham (1969) referred to peaty gleysoils with thin Fe bands in Malaysia. Podzols with strong hydromorphic properties were described from the northern Andes and Costa Rica by Sevink (1984), Hetsch & Hoheisel (1976) and Grieve et al. (1990). The development of more acidic soil conditions, deeper ectoorganic horizons and increased translocation of sesquioxides in the UMCF above 2100 m is apparently due to the pronounced changes in the climatic water balance. Transition from the UMCF to the SCF (2600 to 3400 m) is characterized by increasing dominance of hydromorphic processes over podzolization. A detailed description of the hypsometric changes in soil properties in the study area is provided by Schawe et al. (2007).

Nutrient input. Data on atmospheric inputs of nutrients in tropical forests are presented by Bruijnzeel (1998, 1991), Hafkenscheid (2000) and Proctor (2005). There is great variation in atmospheric nutrient input due to varying amounts and seasonality of rainfall, site effects, and long distance influence of burning activities in the Amazonian lowlands and air mass transport from the altiplano of the Andes (Kellner 2006). It can be assumed that increase of rainfall with altitude leads to higher nutrient inputs in the upper vegetation belts at generally low nutrient concentration in incident rainfall. Concentrations of main nutrient elements are within the range of those reported for the tropical Andes, but altitudinal sequences are rare (Clark et al. 1998, Hafkenscheid 2000, Veneklaas 1990, Wilcke et al. 2001). In the Colombian Andes, nutrient input by rainfall decreases with elevation due to decreasing precipitation (Veneklaas 1990). In the study area, however, input of the main cations increases with elevation despite a decrease of concentration (Table 6). Since total annual rainfall and rainfall nutrient input may vary considerably, measurements over periods of several years are necessary (Bendix et al. 2004). The enrichment of nutrient concentration measured in throughfall in LMF is similar to that in other tropical studies, except for nitrate (Table 6). In UMCF and SCF we found a strong decrease for nitrate from incident rainfall to throughfall. It has been suggested that high epiphytic biomass in tropical cloud forests act as "nutrient capacitor", absorbing sulphur (SO4) and nitrogen (N) (Cavelier et al. 1997). The decrease of nitrate and sulphate with throughfall in the Yungas might support this hypothesis.

Table 7. Nutrient input with throughfall and litter (kg ha^{-1} a^{-1}) in tropical montane forests.

country and forest type	N	P	Ca	Mg	K
Bolivia LMF	229	31	138	52	67
Bolivia UMCF	64	4	48	15	20
Bolivia SCF	51	4	52	14	18
Colombia UMCF[1]	44	3	-	-	78
Ecuador LMF[2]	221	20	175	79	199
World LMF[3]	31-136	1-9	23-154	11-37	68-154
World UMCF[3]	36-117	1-6	14-85	10-30	69-128

The main contribution to nutrient input in the montane forests of the study area is provided by cycling of litter (Table 4). Compared to cloud forests world-wide (Hafkenscheid 2000), litter amounts in the study area are within the range of variation. Litter nutrient input is about 70% higher in LMF than in UMCF and SCF, mainly due to 3–4 times higher litter fall rates. Total nutrient input with throughfall and litter range at the lower end compared to Colombia (UMCF) and Ecuador (LMF/UMCF) (Table 7). The highest decrease of nutrient input by litter from UMCF to SCF is in N, P and S, indicating a decreasing availability of these nutrients with altitude. The maximum difference in mineralization rate (Oi-horizon) between LMF and SCF exists for N (SCF: k_{oi} = 0.86 leaf litter) and P (SCF: k_{oi} = 1.05 leaf litter) (Schawe 2005). Low turnover rates for nitrogen are influenced by water saturation in the upper soil horizon. Cavelier et al. (2000) found decreasing N-mineralization with increasing soil water content in montane rainforests of Colombia.

Conclusions

Soil formation in the Bolivian Yungas is controlled by the decrease in temperature with altitude and increase in precipitation, accompanied by a decrease in evapora-tion. The degree of soil moisture and soil waterlogging in the Ah- and O-horizons increases from LMF to UMCF, and stagnic conditions are present permanently in SCF, causing hydromorphy. Podzolization and hydromorphy are the principal soil forming processes in these altitudinal belts. The decrease of temperature with altitude and the increase of upper soil waterlogging lead to a significant decrease of mineralization rates in UMCF and SCF. Due to the acidity and very low nutrient availability of soils, root development and root mass are concentrated in the ectoorganic soil layer (Leuschner et al. 2007; see also chapter 8 of this volume). This correlates with an increase of carbon stocks in the ectoorganic horizon; more than half of soil organic carbon is accumulated here in UMCF and SCF. In contrast with other studies in montane cloud forests, however, a continuous rise of total C-stocks with altitude was not found.

Nutrient stocks are concentrated in the ectoorganic layer mainly in the SCF. From LMF to UMCF and SCF mineralization rate decreases. Together with the wide C/N ratio, low pH, high Al-saturation and strong hydromorphy, nutrient availability in the root zone may be hampered. Nutrient input of main cations by throughfall decreased from LMF to the upper vegetation belts and epiphytic biomass may act as "nutrient capacitor" for sulphur and nitrogen. Litter fall is the main source of nutrient input and is much less in UMCF and SCF than in LMF, and is low compared to studies from Jamaica (SCF) and Ecuador (UMCF). N, P and S are the main critical nutrients but more research on nutrient uptake in UMCF and SCF is necessary to support the hypothesis that the reduced tree stature with elevation is caused by nutrient limitation.

Despite the marked differences in short-wave radiation, PPFD and PET between the three investigated forest types, these hydrometeorologic factors do not seem to exhibit a direct influence on forest stature and composition, and probably not on forest productivity either (Homeier 2004). Instead, the effect was presumably indirect. The strong increase in precipitation with elevation induced reduced litter mineralization and nutrient availability, increased nutrient leaching and reduced soil aeration (Bach 2004, Schawe et al. 2007). These factors may be considered to be the principal ones responsible for the observed changes in vegetation structure and plant diversity along the elevational gradient.

Acknowledgments. This study was carried out in close collaboration with the Instituto de Ecología of the Universidad Mayor de San Andrés, La Paz, the National Park Administration (SERNAP) in Bolivia and the Department of Systematic Botany, University of Göttingen. Linguistic correction of the text was done by S.R. Gradstein. The study was funded by the Deutsche Forschungsgemeinschaft (DFG) (Grants GE 431/12–3 to G. Gerold and GR 1588/4 to S.R. Gradstein).

References

Ataroff M (1998) Importance of cloud water in Venezuelan Andean cloud forest water dynamics. In: Schemenauer RS, Bridgeman HA (eds) First Int. conference on fog and fog collection. ICRC Ottawa, pp 25-28

Aylett GP (1985) Irradiance interception, leaf conductance and photosynthesis in Jamaican upper montane rain forest trees. Photosynthetica 19: 323-337

Bach K., Kessler M, Gradstein SR (2007) A simulation approach to determine statistical significance of species turnover peaks in a species-rich tropical cloud forest. Diversity and Distributions 13: 863–870.

Bach K (2004) Vegetationskundliche Untersuchungen zur Hoehenzonierung tropischer Bergregenwälder in den Anden Boliviens. Dissertation, University of Göttingen

Bach K, Schawe M, Beck S, Gerold G, Gradstein SR, Moraes M (2003) Vegetación, suelos y clima en los diferentes pisos altitudinales de un bosque montano de Yungas, Bolivia. Primeros resultados. Ecología en Bolivia 38: 3-14

Barthlott W, Lauer W, Placke A (1996) Global distribution of species diversity in vascular plants: towards a world map of photodiversity. Erdkunde 50: 317-327

Beck S (1988) Las regions ecológicas y las unidades fitogeográficas de Bolivia. In: Morales CB (eds) Manual de Ecologia, La Paz, pp 233-271

Beck S (1998) Floristic inventory of Bolivia – an indispensable contribution to sustainable development. In: Barthlott W, Winiger M (eds) Biodiversity. A challenge for development, research and policy. Springer Berlin, pp 243-266

Bendix J, Rollenbeck R, Palacios W (2004) Cloud detection in the tropics – a suitable tool for climate – ecological studies in the high mountains of Ecuador. International Journa of Remote Sensing 25: 4521-4550

Böhner J, Köthe R, Trachinow C (1997) Weiterentwicklung der automatischen Reliefanalyse auf der Basis von digitalen Geländemodellen. Göttinger Geographische Abhandlungen 100: 3-21

Braun-Blanquet J (1964) Pflanzensoziologie. Springer, Berlin.

Brouwer J (1996) Nutrient cycling in pristine and logged tropical rainforest. A study in Guyana. University of Utrecht, Utrecht

Bruijnzeel LA (1991) Nutrient input-output budgets of tropical forest ecosystems: a review. Journal of Tropical Ecology 7: 1-23

Bruijnzeel LA, Waterloo ML, Proctor J, Kuiters AT, Kotterink B (1993) Hydrological observations in montane rain forests on Gunung Silam, Sabah, Malaysia, with special reference to the 'Massenerhebung' effect. Journal of Ecology 81: 145-167

Bruijnzeel LA, Proctor J (1995) Hydrology and biogeochemistry of tropical montane cloud forests: What do we really know? In: Hamilton LS, Juvik JO, Scatena FN (eds) Tropical Montane Cloud Forests, Ecological studies 110. Springer, New York, pp 38-78

Bruijnzeel LA, Veneklaas EJ (1998) Climatic conditions and tropical montane forest productivity: the fog has not lifted yet. Ecology 79: 3-9

Bruijnzeel LA, Hamilton LS (2000) Decision time for cloud forests. UNESCO, Paris

Cavelier J, Mejía C (1990) Climatic factors and tree stature in the elfin cloud forest of Serranía de Macuira, Colombia. Agricultural and Forest Meteorology 53: 105-123

Bruijnzeel LA (2005) Tropical montane cloud forests: A unique hydrological case. In: Bonell M, Bruijnzeel LA (eds) Forests, water and people in the humid tropics. Cambridge: Cambridge University Press pp 462-483

Cavelier J, Jaramillo M, Solis D, de Leon D (1997) Water balance and nutrient inputs in bulk precipitation in tropical montane cloud forest in Panama. Journal of Hydrology 193: 83-96

Cavelier J, Tanner E, Santamaria J (2000) Effect of water, temperature and fertilizers on soil nitrogen net transformations and tree growth in an elfin cloud forest of Colombia. Journal of Tropical Ecology 16: 83-99

Clark KL, Nadkarni NM, Schaeffer D, Gholz HL (1998) Atmospheric deposition and net retention of ions by the canopy in a tropical montane forest, Monteverde, Costa Rica. Journal of Tropical Ecology 14: 27-45

Dietz J, Hölscher D, Leuschner C, Hendrayanto (2006) Rainfall partitioning in relation to forest structure in differently managed montane forest stands in Central Sulawesi, Indonesia. Forest Ecology and Management 237: 170-178

Doorenbos J, Pruitt WO (1988) Yield response to water. FAO Irrigation and drainage paper 33, Rome

Flenley JR (1996) Cloud forests, the Massenerhebung effect and ultraviolett insolation. In: Hamilton LS, Juvik JO, Scatena FN (eds) Tropical montane cloud forests. Ecological Studies 110. Springer, New York, pp 150-155

Fölster H, Fassbender H (1978) Untersuchungen über Bodenstandorte der humiden Bergwälder in der nördlichen Andenkordillere. Tropentagung "Landnutzung in den humiden Gebieten der Tropen", Göttingen, pp 101-110

Garcia-Nunez C, Azocar A, Rada F (1995) Photosynthetic acclimation to light in juveniles of two cloud forest tree species. Trees 10: 114-124

Gentry AH (1995) Patterns of diversity and floristic composition in neotropical montane forests. In: Churchill SP, Balslev H, Forero E, Luteyn JL (eds) Biodiversity and conservation of neotropical montane forests. The New York Botanical Garden, New York, pp 103-126

Gerold G, Schawe M, Joachim L (2003) Pedoökologische hypsometrische Varianz in ungestörten Bergregenwäldern der Anden (Yungas). Geo-Oeko 24: 153-162

Gerold G (2004) Soil: The foundation of biodiversity. In: Ibisch PL, Mérida G (eds) Biodiversity: The richness of Bolivia. FAN-Santa Cruz, pp 17-31

Gerold G, Schawe M, Bach K (2008) Hydrometeorologic and pedologic patterns in relation to montane forest types along an elavational gradient in the Yungas, Bolivia. Die Erde (accepted)

Gill AM (1969) The ecology of an elfin forest in Puerto Rico. 6. Aerial Roots. Journal of Arnold Arboretum 50: 197-209

Grieve I, Proctor J, Cousins S (1990) Soil variation with altitude on Volcan Barva, Costa Rica. Catena, Supplement 17: 525-534

Grubb PJ (1974) Factors controlling the distribution of forest-types on tropical mountains: New facts and a new perspective. In: Flenley JR (ed) Altitudinal Zonation in Malaysia. Third Aberdeen-Hull Symposium on Malaysian Ecology. University of Hull, Department of Geography, Miscellaneous Series, pp 13-46

Grubb PJ (1977) Control of forest growth and distribution on wet tropical mountains: with special reference to mineral nutrition. Annual Review of Ecology and Systematics 8: 83-107

Hafkenscheid RRLJ (2000) Hydrology and biogeochemistry of tropical montane rain forests of contrasting stature in the Blue Mountains, Jamaica. Dissertation, Free University of Amsterdam

Hamilton LS (1995) Mountain cloud forest conservation and research: A synopsis. Mountain Research and Development 15: 259-266

Heinrichs H, Brumsack H, Loftfield N, König N (1986) Verbessertes Druckaufschlusssystem für biologische und anorganische Materialien. Zeitschrift für Pflanzenernährung und Bodenkunde 149: 350-353

Hetsch W, Hoheisel H (1976) Standorts- und Vegetationsgliederung in einem tropischen Nebelwald. Allgemeine Forst- und Jagdzeitschrift 147: 200-209

Herrmann R (1971) Die zeitliche Änderung der Wasserbindung im Boden unter verschiedenen Vegetationsformationen der Höhenstufen eines tropischen Hochgebirges. Erdkunde 25: 90-112

Holder CD (2003) Fog precipitation in the Sierra de las Minas Reserve, Guatemala. Hydrological Processes 17: 2001-2010

Holwerda F (2005) Water and energy budgets of rain forests along an elevation gradient under maritime tropical conditions. Dissertation, Free University of Amsterdam

Homeier J (2004) Baumdiversität, Waldstruktur und Wachstumsdynamik zweier tropischer Bergregenwälder in Ecuador und Costa Rica. Dissertationes Botanicae 391

Juvik JO, Nullet D (1995) Relationships between rainfall, cloudwater interception and canopy throughfall in a Hawaiian montane forest. In: Hamilton LS, Juvik JO, Scatena FN (eds) Tropical Mountain Cloud Forests. Ecological Studies 110. Springer, New York, pp 165-182

Kappelle M, Brown AD (eds) (2001) Bosques nublados del neotrópico. Santa Domingo de Heredia, Costa Rica

Keig G, Flemming PM, McAlpine JR (1979) Evaporation in Papua New Guinea. Journal of Tropical Geography 48: 19-30

Kellner Th (2006) Niederschlagsnährstoffeinträge im Bergregenwaldökosystem der Yungas (Bolivien). Diplomarbeit, University of Göttingen

Kessler M (2000) Elevational gradients in species richness and endemism of selected plant groups in the central Bolivian Andes. Plant Ecology 149: 181-193

Kessler M, Beck SG (2001) Bolivia. In: Kappelle M, Brown AD (eds) Bosques nublados del néotropico, Santo Domingo de Heredia, Costa Rica, pp 581-622

Kitayama K (1992) An altitudinal transect study of the vegetation on Mount Kinabalu, Borneo. Vegetatio 102: 149- 171

Kitayama K (1995) Biophysical conditions of the montane cloud forests of Mount Kinabalu, Sabah, Malaysia. In: Hamilton LS, Juvik JO, Scatena FM (eds) Tropical montane cloud forests, Ecological studies 110. Springer, New York, pp 115-125

Kitayama K, Mueller-Dombois D (1994) An altitudinal transect analysis of the windward vegetation on Haleakala, a Hawaiian island mountain (2). Vegetation zonation. Phytocoenologia 24: 135-154

Körner C (1982) CO_2 exchange in the alpine sedge *Carex curvula* as influenced by canopy structure, light and temperature. Oecologia 53: 98-104

Körner C, Allison A, Hilscher H (1983) Altitudinal variation of leaf diffusive conductance and leaf anatomy in heliophytes of montane New Guinea and their interrelation with microclimate. Flora 174: 91-135

Lauscher F (1976) Weltweite Typen der Höhenabhängigkeit des Niederschlags. Wetter Leben 28: 89-90

Leuschner Ch, Moser G, Bertsch Ch, Röderstein M, Hertel D (2007) Large altitudinal increase in tree root/shoot ratio in tropical mountain forests of Ecuador. Basic and Allpied Ecology 8: 219-230

Lyford W H (1969) The Ecology of an elfin forest in Puerto Rico. Soil, root, and earthworm relationships. Journal of Arnold Arboretum 50: 210-224

Luer B, Böhmer A (2000) Vergleich zwischen Perkolation und Extraktion mit 1 M NH4Cl-Lösung zur Bestimmung der effektiven Kationenaustauschkapazität (KAKeff.) von Böden. Journal of Plant Nutrition and Soil Science 163: 555-557

Motzer Th (2003) Bestandesklima, Energiehaushalt und Evapotranspiration eines neotropischen Bergregenwaldes. Mannheimer Geographische Arbeiten 56

Myers N, Mittermeier RA, Mittermeier CG, Fonseca GABD, Kent J (2000) Biodiversity hotspots for conservation priorities. Nature 403: 853-858

Odum HT (1970) Rain forest structure and mineral cycling homeostasis. In: Odum HT, Pidgeon RF (eds) A tropical rain forest, a study of irradiation and ecology at El Verde, Puerto Rico, H3-H52. U.S. Atomic Energy Commission, Oak Ridge, Tennessee

Pareja Millan A (2005) Variacion hipsometrica y vertical del clima en el bosque nublado, caso Cerro Hornuni – Parque Nacional Cotapata. Tesis de Maestria, Instituto de Ecologia, La Paz

Pendry CA, Proctor J (1996) The causes of altitudinal zonation of rain forests on Bukit Belalong, Brunei. Journal of Ecology 84: 407-418

Proctor J (2005) Rainforest mineral nutrition: the "black box" and a glimpse inside it. In: Bonell M, Bruijnzeel LA (eds.) Forests, water and people in the humid tropics. Cambridge University Press, Cambridge, UK, pp 422-446

Roman LA, Scatena FN (in press) Tropical montane cloud forest soils: an overview. In: Bruijnzeel S, Juvik J, Hamilton L (eds) Mountains in the mist. University of Hawaii Press, Hawaii

Santiago LS, Goldstein G, Meinzer FC, Fownes JH, Mueller-Dombois D (2000) Transpiration and forest structure in relation to soil waterlogging in a Hawaiian montane cloud forest. Tree Physiology 20: 673-681

Schawe M (2005) Hypsometrischer Klima- und Bodenwandel in Bergregenwaldökosystemen Boliviens. Dissertation Göttingen

Schawe M, Glatzel S, Gerold G (2007) Soil development along an altitudinal transect in a Bolivian tropical montane rainforest: Podzolization vs. hydromorphy. Catena 69: 83-90

Schlichting E, Blume H, Stahr K (1995) Bodenkundliches Praktikum. Berlin

Schrumpf M, Guggenberger G, Valarezo C (2001) Tropical montane rainforest soils. Die Erde 132: 43-59

Sevink J (1984) An altitudinal sequence of soils in the Sierra Nevada de Santa Marta (Ecoandes). In: Van der Hammen T, Ruiz PM (eds) Studies on tropical Andean Ecosystems Vol. 2. Cramer, Vaduz, pp 131-138

Silver W, Lugo AE, Keller M (1999) Soil oxygen availability and biogeochemistry along rainfall and topographic gradients in upland wet tropical forest soils. Biogeochemistry 44: 301-328

Soil Survey Staff (2003) Keys to soil taxonomy. USDA-NRCS Agric. Handbook, Washington D.C.

Stadtmüller T (1987) Cloud forests in the humid tropics. A bibliographic review. The United Nations University, Tokyo

Tanner W, Beevers H (1990) Does transpiration have an essential function in long- distance ion transport in plants? Plant Cell Environment 13: 745-750

Tanner E, Vitousek P, Cuevas E (1998) Experimental investigation of nutrient limitation of forest growth on wet tropical mountains. Ecology 79: 10-22

Van der Hammen T, Mueller-Dombois D, Little MA (eds) (1989) Manual of methods for mountain transect studies. Comparative studies of tropical mountain ecosystems. International Union of Biological Sciences, Paris

Veneklaas EJ, van Ek R (1990) Rainfall interception in two tropical montane rain forests, Colombia. Hydrological processes 4: 311-326

Vis M (1986) Interception, drop size distributions and rainfall kinetic energy in four Colombian forest ecosystems. Earth Surface Processes and Landforms 11: 591-570

Weischet W (1969) Klimatologische Regeln zur Vertikalverteilung der Niederschläge in Tropengebirgen. Die Erde 100: 287-306

Whitmore T, Burnham C (1984) Tropical Rain Forests of the Far East. Oxford University Press, New York.

Wilcke W, Yasin S, Valarezo C, Zech W (2001) Change in water quality during the passage through a tropical montane rain forest in Ecuador. Biogeochemistry 55: 45-72

Wilcke W, Yasin S, Abramowski U, Valarezo C, Zech W (2002) Nutrient storage and turnover in organic layers under tropical montane rain forest in Ecuador. European Journal of Soil Science 53: 15-27

Yasin S (2001) Water and nutrient dynamics in microcatchments under montane forest in the South Ecuadorian Andes. Bayreuther Bodenkundliche Berichte 73

Biodiversity and Ecology Series (2008) 2: 163-176
The Tropical Mountain Forest – Patterns and Processes in a Biodiversity Hotspot
edited by S.R. Gradstein, J. Homeier and D. Gansert
Göttingen Centre for Biodiversity and Ecology

Indigenous land use practices and biodiversity conservation in southern Ecuador

Perdita Pohle

Institute of Geography, Chair for Human Geography and Development Studies, Friedrich-Alexander-Universität Erlangen-Nürnberg, Kochstr. 4/4, 91054 Erlangen, Germany,
ppohle@geographie.uni-erlangen.de

Abstract. It is well understood that any attempt to preserve primary forest in the tropics is destined to fail if the interests and use claims of the local population are not taken into account. Therefore, in addition to strict protection of the forests, an integrated concept of nature conservation and sustainable land use development is being sought. The DFG-research project presented here will figure out the extent to which traditional ecological knowledge and biodiversity management strategies can be made available for long-term land use development. Ethno-ecological and agro-geographical research methods were used to document indigenous knowledge of traditionally utilized wild and domestic plants, to analyze current forms of land use, and to evaluate ethno-specific life-support strategies and strategies of biodiversity management among Shuar and Saraguro communities in southern Ecuador.

Introduction

Loss of biodiversity and land degradation are not environmental problems as they are usually labeled, but problems created by the culture-specific relationship between people or societies and their natural environment - an inventive thought, which was first formulated by Paul Messerli (1994, p. 144), a geographer from Bern. In current research programmes with the objective of preserving ecosystems and habitats, it is essential not only to consider the natural ecosystem but also to include human dimensions. Centre of attention for such research must be human actions: the interplay of acting individuals (actors) and social groups (collectives of actors, e.g. communities) under specific social conditions. In the agricultural frontier zone of southern Ecuador, a region of heterogenic ethnic, socio-cultural and socio-economic structures, profound knowledge of ethnic-specific human ecological parameters is crucial for the sustainable utilization and conservation of tropical mountain forests. Today it is well understood that any attempt to preserve primary forest in the tropics is destined to fail if the interests and use claims of the local population are not at the same time, and in the long term, taken into account. Therefore, in addition to strict protection of the forests, an integrated concept of nature conservation and sustainable land use development needs to be sought (e.g. Ellenberg 1993).

Goal of the DFG-research project presented here is to determine the extent to which traditional ecological knowledge and indigenous biodiversity management strategies can be made available for long-term land use development.

Research area and indigenous groups

The tropical mountain rainforests of the eastern Andean slopes in southern Ecuador have been identified as a one of the so-called "hot spots" of biodiversity worldwide (Barthlott et al. 1996, Myers et al. 2000). These mountain forest ecosystems, which have been described as particularly sensitive (cf. Die Erde 2001), have come under enormous pressure from the expansion of agricultural land (esp. pastures), the extraction of timber, the mining of minerals, the tapping of water resources and other forms of human intervention.

The area under study comprises the northern and eastern buffer zones of Podocarpus National Park, settled by indigenous Saraguro and Shuar communities as well as Mestizo-Colonos. Although the Mestizo-Colonos are by far the most prominent ethnic group in numbers, this article will focus on the indigenous Saraguro and Shuar communities settling around Podocarpus National Park (Fig. 1). The Shuar area of settlement extends from the lower levels of the tropical mountain rainforest (approx. 1400 m) down to the Amazonian lowland (Oriente) in the region bordering Peru. The Shuar belong to the Jívaro linguistic group (Amazonian Indians). They are typical forest dwellers who practice shifting cultivation in subsistence economy. Besides, they fish, hunt and gather forest products. During the last decades some Shuar have also begun to raise cattle and some are engaged in timber extraction as well. They are typical forest dwellers who practice shifting cultivation in subsistence economy. Besides, they fish, hunt and gather forest products. During the last decades some Shuar have also begun to raise cattle and some are engaged in timber extraction as well.

The Saraguros are traditionally Quichua-speaking highland Indians who live as agro-pasturalists for the most part in the temperate mid-altitudes (1800–2800 m) of the Andes (Sierra) in southern Ecuador. As early as the 19th century, the Saraguros kept cattle to supplement their traditional system of mixed cultivation, featuring maize, beans, potatoes and other tubers (Gräf 1990). By now, cattle ranching has developed as the main branch of their economy.

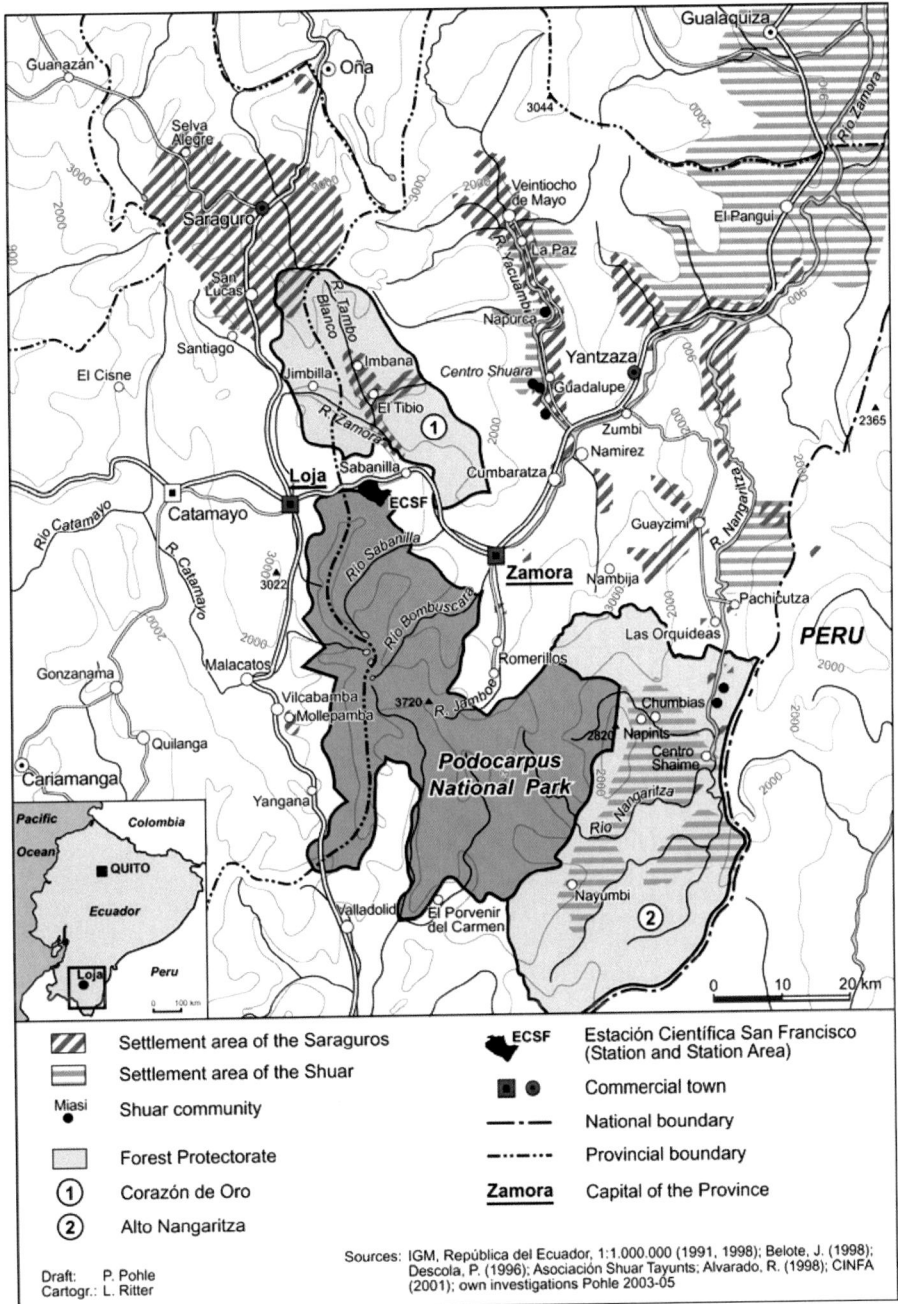

Figure 1. The Podocarpus National Park and settlement areas of indigenous groups.

Aims and methods of the ethnoecological research project

The project was carried out within the DFG Research Unit FOR 402: "Functionality in a Tropical Mountain Rainforest" (www.bergregenwald.de). Within the ethnoecological project, ethnicity is viewed as a driving factor in the relationship between man and his environment. Fundamental differences between the indigenous Shuar and Saraguro communities as well as the local Mestizo-Colonos occur not only in attitudes towards the tropical rainforest and the management of forest resources, but also in wider economic and social activities, including all strategies for maintaining livelihood (Pohle & Gerique 2006).

During 2004, 2005 and 2006 ethnoecological, especially ethnobotanical and agrogeographical research was undertaken in sample communities of the Shuar (Shaime, Napints, Chumbias) and the Saraguros (El Tibio). The aims were: (1) to document the indigenous and local knowledge of traditionally utilized wild and cultivated plants (the ethnobotanical inventory was undertaken according to the "Code of Ethics"); (2) to analyze current forms of land use including the cultivation of forest and home gardens; and (3) to evaluate ethno-specific life-support strategies as well as strategies for natural resource management.

Significance of plant use for the Shuar and Saraguro communities

In biodiversity-rich places local people usually have a detailed ecological knowledge, e.g. of species, ecosystems, ecological relationships and historical or recent changes to them (e.g. Warren et al. 1995). This applies wholly to the Shuar communities. As traditional forest dwellers the Shuar of the Nangaritza valley have a comprehensive knowledge of plants and their utilization. All households make extensive use of forest products.

In Fig. 2 the number of plant species (wild and cultivated) used by the Shuar and Saraguro communities are listed. Plant uses were recorded according to categories of utilization like food, medicine, construction etc. According to the ethnobotanical survey[1], the actual inventory of traditionally used wild plants of the Shuar includes 211 species. Most of the plants are used to supplement the diet (74). Given the lack of state health care, medicinal plants also assume great significance (63). Furthermore, many plants are used as construction material (67), as tools and for handicrafts (37), as fuel, fodder or as ritual plants. The Shuar use forest products exclusively for their own needs and there is virtually no commercialization.

---

[1] The ethnobotanical survey was conducted by Andrés Gerique.

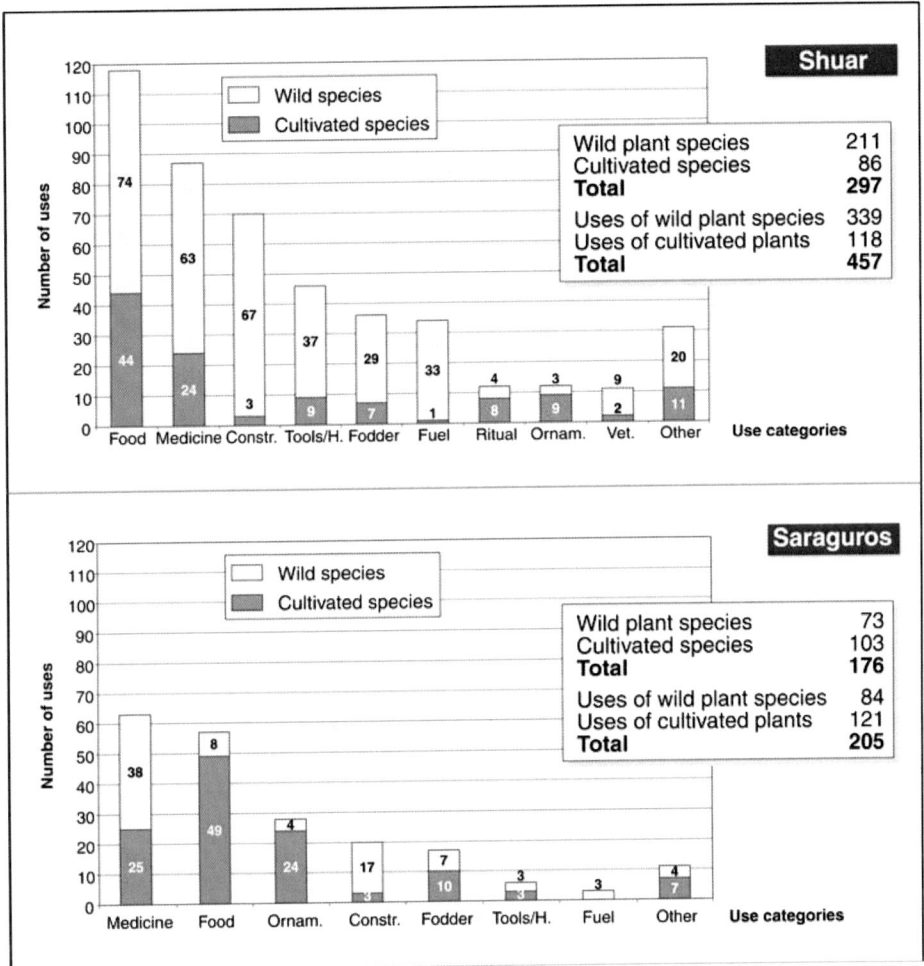

Figure 2. Wild and cultivated plant species used by the Shuar (Shaime, Chumbias, Napints) and the Saraguros (El Tibio) according to use categories. Note: one species can be found in more than one use category.

   The Saraguros from El Tibio have a far less comprehensive knowledge on wild plant species and their utilization. The actual ethnobotanical inventory includes only 73 wild plant species. Most of them are ruderal plants used as medicine (38) or plants used for their wood (17). As agro-pasturalists they have conversed most of the primary forest into pastures, home gardens and fields, leaving forest remains only along mountain ridges or in river ravines. Their actual plant knowledge reflects this traditional way of life. They have a comprehensive knowledge of cultivated plants (103), mainly pasture and crop plant species - even more than the Shuar (86) - but they are less familiar with forest plant species. The latter knowledge is mainly reduced to woody varieties which they extract and sell before clearing the forest.

Agrobiodiversity in Shuar and Saraguro tropical home gardens (huertas)

The results of the ethnobotanical and agrogeographical studies underline the thesis that the tropical home gardens of indigenous communities are places of high agrobiodiversity and refuges of genetic resources (Watson & Eyzaguirre 2002). Furthermore, they contribute significantly to securing and diversifying food supplies.

The forest gardens of the Shuar (Fig. 3) are characterized by an especially great diversity of species and breeds. In five huertas studied (size: approx. 600–1000 m²), a total of 185 wild and cultivated plant species and breeds were registered. For the most part they serve as nutritional items (58%) or medicines (22%). The main products cultivated are manioc (*Manihot esculenta*) and taro (*Colocasia esculenta*), along with various breeds of plantains (*Musa* sp.). Moreover, the planting of a large number of traditional local breeds was documented: e.g. 29 breeds of manioc and 21 breeds of *Musa* sp. – a further indication of the crucial significance that home gardens have for the in situ conservation of botanical genetic resources (Münzel 1989, p. 434).

The huertas of the Saraguros likewise display a great diversity of useful plants. In one sample home garden studied in El Tibio (Fig. 4), 51 species of cultivated plants were identified. Again, the majority are plants that supply nutritional value (41%), followed by medicinal and ornamental plants (each 20%). The most important cultivated products are plantains, tubers and various types of fruit. Given their relatively dense and tall stands of trees, the multi-tiered arrangement of plants and the great diversity of species, the gardens of the Saraguros can be seen as an optimal form of exploitation in the region of tropical mountain rainforests.

Indigenous concepts of biodiversity management - their contribution to a sustainable land use development

If the hypothesis is accepted that a multi-facetted economic and cultural interest in the forest on the part of indigenous and local communities offers effective protection against destruction, then a key role must be assigned to the analysis and evaluation of the ethno-specific knowledge about tropical mountain rainforests and their potential uses. Both indigenous groups have developed natural resource management strategies that could be used and expanded, in line with the concept "preservation through use", for future biodiversity management, but this should be done only in an ethno-specific way.

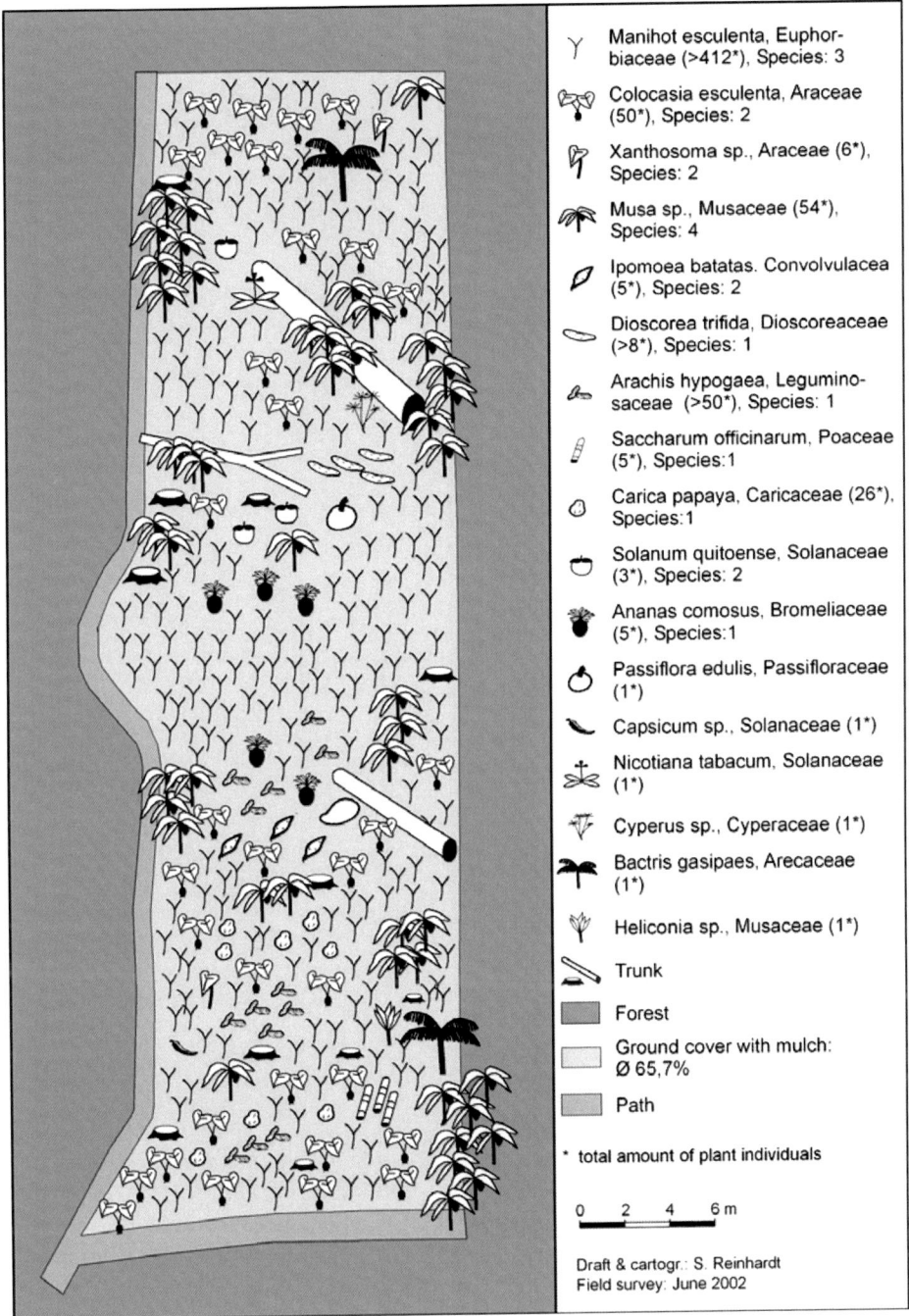

Figure 3. Shuar forest garden (huerta) in the Nanagarita valley.

Saraguro home garden in El Tibio (southern Ecuador) 1770 m

Way to El Tibio alto

Field (chacra)

Yard

Veranda

Residential building

Kitchen

Pond

Washbasin

Field (chacra)

Field (chacra)

0   5   10 m

Draft: P. Pohle, A. Tutillo, E. Tapia
Field survey: September 2003
Cartogr.: L. Ritter

Legend:
- Trail
- Trail / gutter
- Fence
- Grove of plantains (*Musa sp.*, 4 breeds) "number of stems"
- Tree
- Bush
- Tree stump
- Shelter for pigs
- Firewood
- Litter
- Latrine

	Spanish name	Species	Family		Spanish name	Species	Family
A	Azucena	*Lilium candidum*	Liliaceae	HL	Hierba Luisa	*Cymbopogon citratus*	Poaceae
Ach	Achira	*Canna indica*	Cannaceae	Hm	Hierba morocho	*Sporobulus indicus*	Poaceae
Ag	Aguacate	*Persea americana*	Lauraceae			*Iochroma sp.*	
Ap	Aguacate pequeñito	*Persea sp.*	Lauraceae	K	Café	*Coffea arabica*	Rubiaceae
Be	Begonia	*Begonia tuberosa*	Begoniaceae	L	Luma	*Pouteria lucuma*	Sapotaceae
C	(Pasto) Cariamanga	*Tripsacum sp.*	Poaceae	M	(Pasto) Merkeron	*Setaria sphacelata*	Poaceae
Ca	Cabuya	*Fourcraea gigantea*	Agavaceae	Ma	Matico	*Piper aduncum*	Piperaceae
Cd	Chadan	(Medicinal plant)	Meliaceae	Mg	Malva goma	*Malva sp.*	Malvaceae
Ce	Cedro	*Cedrela montana*	Meliaceae	Mn	Mandarina	*Citrus reticulata*	Rutaceae
Ch	Chimboro	*Cestrum sendtnerianum*	Solanaceae	Mo	Mortiño	*Solanum americanum*	Solanaceae
Cha	Chabela	*Impatiens balsaminea*	Balsaminaceae	N	Naranjilla	*Solanum quitoense*	Solanaceae
Chi	Chilco blanco	*Baccharis sp.*	Asteraceae	Na	Naranjo	*Citrus sinensis*	Rutaceae
Cm	Camote	*Ipomoea batatas*	Convolvulaceae	Ni	Níspero	*Eriobotrya japonica*	Rosaceae
Cñ	Caña agria	*Costus comosus*	Zingiberaceae	P	Palmillo	*Crocosmia hybr.*	Iridaceae
Cp	Capulí			Pch	Papa china	*Colocasia esculenta*	Araceae
Cr	Corulla	(Fodder plant)		Pd	Pedorrera	*Ageratum conyzoides*	Asteraceae
D	Durazno	*Prunus persica*	Rosaceae	Pe	Pepino	*Cucumis sativus*	Cucurbitaceae
E	(Pasto) Elefante	*Pennisetum purpureum*	Poaceae	Pel	Pelma	*Xanthosoma sp.*	Araceae
G	(Pasto) Gramalote	*Axonopus scoparius*	Poaceae	Pg	Pigllo	*Euphorbia laurifolia*	Euphorbiaceae
Gb	Guabo	*Inga sp.*	Mimosaceae	Pi	Piña	*Ananas comosus*	Bromeliaceae
Gr	Granadilla	*Passiflora ligularis*	Passifloraceae	Pm	Palma de mayo	*Yucca elephantipes*	Agavaceae
Gt	Guato	*Erythrina edulis*	Fabaceae	Pn	Pino	*Pinus radiata*	Pinaceae
Gu	Guizho	*Heliopsis canescens*	Asteraceae	S	Sango	*Xanthosoma sagitifolium*	Araceae
Gy	Guayaba	*Psidium guajava*	Myrtaceae	Sh	Shiran		
H	Hortensia	*Hydrangea hortensis*	Hydrangeaceae	T	Tomate de árbol	*Cyphomandra betacea*	Solanaceae
Hi	Higo	*Ficus sp.*	Moraceae	Z	Cebolla larga	*Allium fistulosum*	Liliaceae

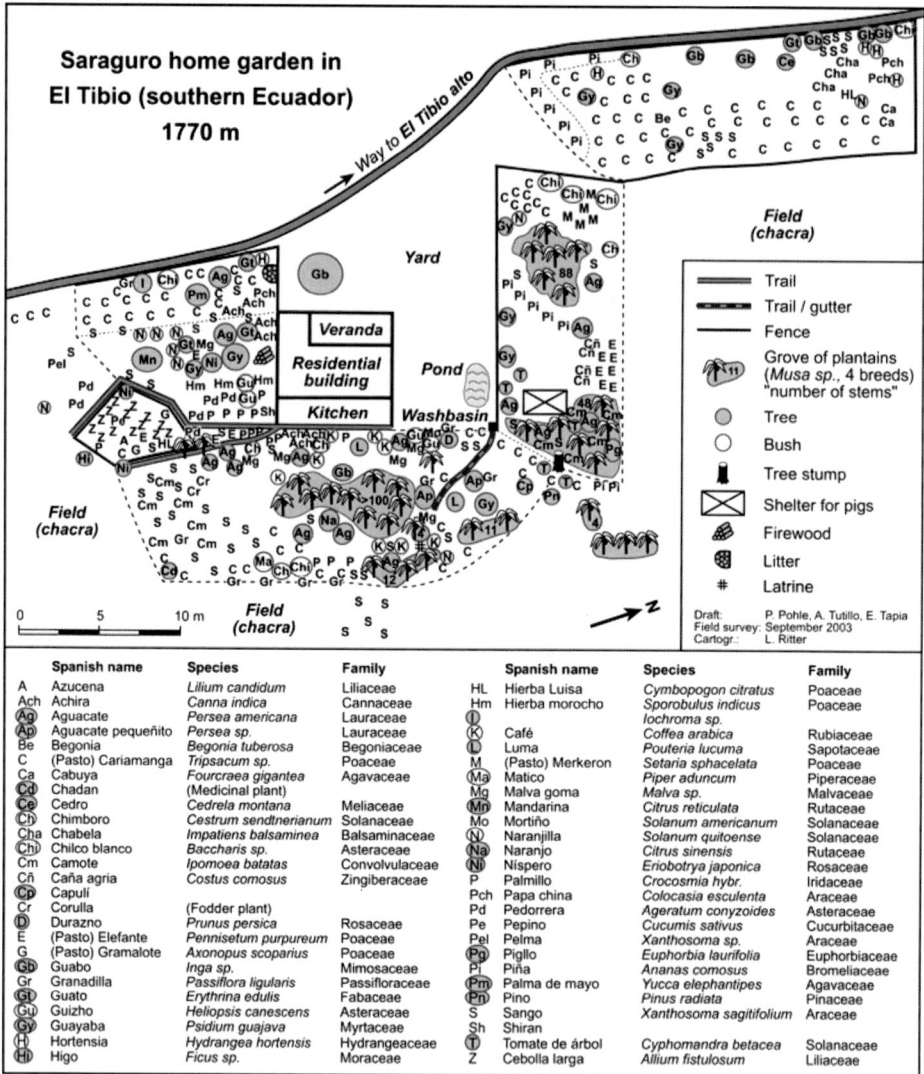

Figure 4. Saraguro home garden in El Tibio (1770 m).

The Shuar traditional way of managing biodiversity is based on a sense of being closely bound culturally, spiritually and economically to the forest. The sustainability of their form of land use has long since been put to the test (Münzel 1977, 1987). As traditional forest-dwellers, sustainable elements of biodiversity management can be found in (Fig. 5):

(1) Their regulated practice of shifting cultivation, which - given the correspondingly long time for regeneration - is thought to conserve the

Figure 5. Napints (1000 m): scattered settlement of the Shuar in the tropical rainforest at the eastern periphery of Podocarpus National Park. Photograph A. Gerique.

soil and the vegetation. The system of cultivation and fallow on small rotating plots (Fig. 6) has much in common with ecological succession in that it uses the successional process to restore the soil and the vegetation after use for farming (Kricher 1997, p. 179). In the Shuar forest gardens the fallow periods last for about 24–30 years while the cultivation periods cover 4 years.

(2) Their tending of forest gardens according to principles of agroforestry and mixed cropping with a high agrobiodiversity and a particular high breed variety of cultivated plants. As it is commonly known, polycultures are more resistant to insect attacks and plant diseases.

(3) The natural fertilization of soils by mulching and the use of digging sticks and dibbles as a suitable form of cultivating the soil.

(4) Their sustainable use of a broad spectrum of wild plants in small quantities, satisfying only subsistence needs and avoiding over-harvesting.

If the Shuar's traditionally practiced and clearly sustainable plant diversity management is to be preserved, this is possible only by legalization of their territorial claims and comprehensive protection of their territories, for example by demarcating reservations. This appears to be underway with the establishment of a so-called Reserva Shuar (Neill 2005). Additionally, it is necessary to respect and support the Shuar's cultural identity, not only to avoid the loss of traditional environmental knowledge, in particular traditional plant lore. To improve

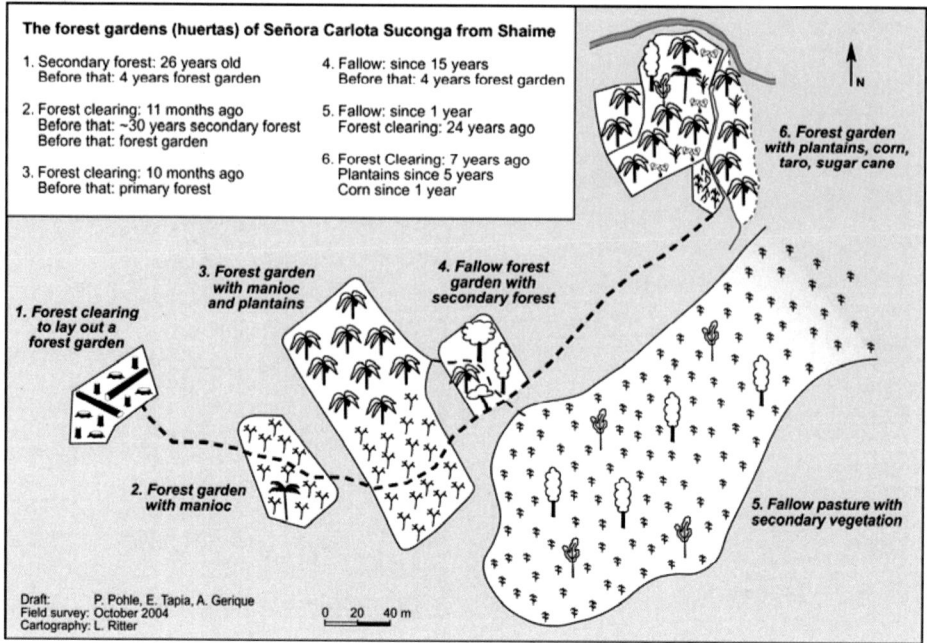

Figure 6. The forest gardens (huertas) of Señora Carlota Suconga from Shaime (920 m).

livelihood in an economic sense, additional sources of financial income are essential. In this line the promotion of ecotourism, support of traditional handicrafts and the cultivation of useful plants for a regional market could be discussed.

While the Shuar's forest management can be evaluated as preserving plant diversity, the sustainability of the Saraguros' use of the environment has yet to be rated. Market-oriented stockbreeding has particularly led in recent decades to the rapid increase of pastures at the expense of forest. In spite of ecological conditions unfavourable to agricultural pursuits (steep V-shaped valleys, acidic soils, extremely high precipitation), these Andean mountain farmers have at least, by means of their intensive form of pasture management, succeeded in generating a sufficiently stable agrarian and cultural landscape (Fig. 7). In contrast to many completely deforested and ecologically devastated areas settled by Mestizo-Colonos (Fig. 8), the richly chequered agrarian landscape of the Saraguros presents, not only esthetically but also ecologically, a fundamentally more positive picture.

Among the Saraguros, initial attempts have also been elaborated to manage biodiversity in line with the concept "preservation through use". The first thing to be mentioned in this context is the keeping of home gardens with a wide spectrum of wild and cultivated plants, particularly woody species. With regard to the diversity of species, the remnants of forest still largely preserved in ecologically unfavourable locations are significant. In order to stem the further loss of

Figure 7. Richly chequered cultural landscape of the Saraguros on a steep slope of the Río Tibio valley with the scattered settlement of El Tibio (1770 m).

biodiversity, however, it will be necessary to convince the Saraguros that in particular scrub- and wasteland (matorral) should be replanted with native tree species. The pressure on the tropical mountain forests caused by the pasturing economy will only be reduced, though, when the Saraguros can be shown a profitable alternative to it. As examples of promising endeavours in this context may be regarded:

(1) The selective timber production and replanting with native tree species as proposed by foresters (Günter et al. 2004).
(2) The introduction of silvipastoral or agroforestry systems.
(3) The market-oriented gardening.
(4) The cultivation and marketing of useful plants, e.g. medicinal herbs.
(5) The promotion of "off-farm" employment opportunities.
(6) The payment for environmental services to protect the watershed area of Loja.

Protecting biological diversity - from National Park to Biosphere Reserve

Podocarpus National Park, covering a total of 146,280 ha, was established in 1982 as southern Ecuador's first conservation area, whose goal is to protect one of the

Figure 8. Deforested and overused agrarian landscape of the Mestizo-Colonos north of Loja.

country's last intact mountain rainforest ecosystems, one particularly rich in species and largely untouched by humans. The creation of a national park in the middle of a fairly densely populated mountain region necessarily gave rise to numerous conflicts of interest and use rights, e.g. agrarian colonization, illegal timber extraction, conflicts about landownership and possession, mining activities, tourism.

The experience in international nature conservation during the past decades has shown that resource management, if it is to be sustainable, must serve the goals of both nature conservation and the use claims of the local population. The strategy is one of "protection by use" instead of "protection from use", a concept that has emerged throughout the tropics under the philosophy "use it or loose it" (Janzen 1992, 1994). In the following, strategies are presented that show a way how people can benefit from the national park without degrading the area ecologically: the implementation of extractive reserves, the promotion of ecotourism and as the most promising approach the establishment of a Biosphere Reserve.

Given the high biodiversity of tropical rainforests and the fact that indigenous people in general have a comprehensive knowledge of forest plants and their utilization, it seems possible according to Kricher (1997, p. 357) to view the rainforest as a renewable, sustainable resource from which various useful products can be extracted on a continuous basis. In view of the high number of plant species that are currently collected by extractivists in the surroundings of Podocarpus National Park (e.g. the Shuar), the preservation of large areas of rainforest would make economic sense as well as serve the interests of conservation and preservation of bio-diversity. Thus, the establishment of an extractive reserve could be suggested as an alternative to deforestation.

In line with the concept "protection by use", ecotourism can be structured such that it is compatible with conservation interests and serves the local economy as well. This is also the experience around Podocarpus National Park. The attraction of the park is clearly the tropical rainforest with its specific wildlife, particularly tropical birds; fewer visitors have botanical or eco-geographical interests. However, compared to other national parks of South America (e.g. Manu National Park of Peru), southern Ecuador and Podocarpus National Park are a major tourist destinations.

The most promising approach, in which conservational protection and sustainable development are the guiding principles, is the integrated concept of conservation and development exemplified by UNESCO's Biosphere Reserve (UNESCO 1984). The idea behind it is to mark out representative sections of the landscape composed of, on the one hand, natural ecosystems (core area) and, on the other, areas that bear the impress of human activity (buffer- and development zone; Erdmann 1996). In Ecuador, three biosphere reserves have already been drawn up (Ministerio del Ambiente 2003). The establishment of such a reserve would also be desirable for southern Ecuador and was recently accepted by UNESCO, in September 2007.

Biosphere reserves are strongly rooted in cultural contexts and traditional ways of life, land use practices and local knowledge and know-how. In the buffer- and development zone of the Podocarpus National Park measures to be taken could rely on the rich ethno-specific traditions in forest- and land use practices by indigenous and local communities. Under the umbrella of an UNESCO Biosphere Reserve in southern Ecuador, not only would the protection of tropical mountain ecosystems be guaranteed, but also the development of ecologically and economically sustainable and socially acceptable forms of land use would be assured.

References

Barthlott W, Lauer W, Placke A (1996) Global distribution of species diversity in vascular plants: towards a world map of phytodiversity. Erdkunde 50: 317-327

Belote JD (1998) Los Sarguros del Sur del Ecuador. Ediciones Abya-Yala, Quito

Die Erde (2001) Themenheft "Tropische Wald-Ökosysteme". Die Erde 132 (1)

Ellenberg L (1993) Naturschutz und Technische Zusammenarbeit. Geographische Rundschau 45: 290-300

Erdmann KH (1996) Der Beitrag der Biosphärenreservate zu Schutz, Pflege und Entwicklung von Natur- und Kulturlandschaften in Deutschland. In: Kastenholz HG, Erdmann KH, Wolff M (eds) Nachhaltige Entwicklung. Zukunftschancen für Mensch und Umwelt. Berlin

Gräf M (1990) Endogener und gelenkter Kulturwandel in ausgewählten indianischen Gemeinden des Hochlandes von Ecuador. München

Günter S, Stimm B, Weber M (2003) Silvicultural contributions towards sustainable management and conservation of forest genetic resources in southern Ecuador. In: Proc. of the 2nd Congress "Conservation of the biodiversity in the Andes and the Amazon region", Loja 25th to 30th of August 2003

Janzen DH (1992) A south-north perspective on science in the management, use and economic development of biodiversity. In: Sandlund OT, Hindar K, Brown AHD (eds) Conservation of Biodiversity for Sustainable Development. Oslo, pp 27-52

Janzen DH (1994) Wildland biodiversity management in the tropics: where are we now and where are we going? Vida Silvestre Neotropical 3: 3-15

Kricher J (1997) A Neotropical Companion. Princeton

Messerli P (1994) Nachhaltige Naturnutzung: Diskussionsstand und Versuch einer Bilanz. In: Bätzing W, Wanner H (eds) Nachhaltige Naturnutzung im Spannungsfeld zwischen komplexer Naturdynamik und gesellschaftlicher Komplexität. Geographica Bernensia Band P30, Bern

Ministerio del Ambiente (2003) Ponencias del Ministerio del Ambiente para el Fortalecimiento y Consolidación del Sistema Nacional De Áreas Protegidas. Primer Congreso Nacional de Áreas Naturales Protegidas. Quito

Münzel M (1977) Jívaro-Indianer in Südamerika. Roter Faden zur Ausstellung, 4. Museum für Völkerkunde. Frankfurt

Münzel M (1987) Kulturökologie, Ethnoökologie und Ethnodesarrollo im Amazonasgebiet. Entwicklungsperspektiven 29. Kassel

Münzel M (1989) Bemerkungen zum indianischen Umweltbewußtsein im Amazonasgebiet. Geographische Rundschau 41: 431-435

Myers N, Mittermeier RA, Mittermeier CG, da Fonseca GAB, Kent J (2000) Biodiversity hotspots for conservation priorities. Nature 403: 853-858

Neill D (2005) Cordillera del Cóndor. Botanical treasures between the Andes and the Amazon. Plant Talk 41: 17-21

Pohle P (2004) Erhaltung von Biodiversität in den Anden Südecuadors. Geographische Rundschau 56: 14-21

Pohle P, Gerique A (2006) Traditional ecological knowledge and biodiversity management in the Andes of southern Ecuador. Geographica Helvetica 4: 275-285

UNESCO (1984) Action plan for biosphere reserves. Nature and Resources 20: 11-22

Warren DM, Slikkerveer LJ, Brokensha G (1995) The cultural dimension of development: Indigenous knowledge systems. London

Watson JW, Eyzaguirre PB (2002) Home gardens and in situ conservation of plant genetic resources in farming systems. Proceedings of the 2. International Home Gardens Workshop, 17-19 July 2001, Witzenhausen, Germany. IPGRI Publications

Biodiversity and Ecology Series (2008) 2: 177-193
The Tropical Mountain Forest – Patterns and Processes in a Biodiversity Hotspot
edited by S.R. Gradstein, J. Homeier and D. Gansert
Göttingen Centre for Biodiversity and Ecology

Sustainable management of tropical mountain forests in Ecuador

Reinhard Mosandl and Sven Günter

Institute of Silviculture, Technical University München, Am Hochanger 13, 85354, Freising, Germany, mosandl@forst.tu-muenchen.de

Abstract. The extent of forest resources and biodiversity are two major elements of the FAO definition of sustainability. As deforestation rate and, presumably, loss of biodiversity in Ecuador are the highest in South America, land use in this country is not sustainable. Our research in a megadiverse mountain rainforest ecosystem in the Andes of South Ecuador demonstrates the feasibility of forest and biodiversity conservation by means of scientifically based forest management. Natural succession may in some cases enhance the return of agricultural land to a more natural state. Reforestation of degraded pastures and enrichment planting with native species lead to secondary forests which can be used for timber production, releasing the pressure from the pristine forests. Single measures cannot lead to sustainable land use but concerted actions may be successful in combating the deforestation process in Ecuadorian tropical mountain forests.

Introduction

The term "Forest Sustainability" was presumably used for the first time in German forestry although the idea of sustainable forest management also developed in other regions of Europe, especially in France during the 16th and 17th century (Schanz 1996, Grober 2000). The oldest known documented application is found in "Sylvicultura oeconomica" of 1713 by von Carlowitz (Hasel 1985, Schanz 1996, Weber-Blaschke et al. 2005). The author coined the term "sustainable use" because he considered the traditional expression "careful use" inadequate for his concept of long term forest management in a wise manner (Grober 2000).

At the 1992 environmental world summit in Rio de Janeiro, "sustainability" was a key word in political discussions. The term "sustainable forest management" was used in the "Forest Principles"[1] and in Chapter 11 of Agenda 21, which were prominent outputs of the United Nations Conference on Environment and De-

---

[1] The full name is the Non-Legally Binding Authoritative Statement on Principles for a Global Consensus on the Management, Conservation and Sustainable Development of All Types of Forest (FAO 2001)

velopment (UNCED) in June 1992 (FAO 2006). The idea of the Forest Principles was to contribute to the management, conservation and sustainable development of all types of forests and to provide for their multiple and complementary functions and uses (FAO 2006). Principle 2b specifically states: "Forest resources and forest lands should be sustainably managed to meet the social, economic, cultural and spiritual needs of present and future generations" (FAO 2006).

Since 1992 the concept of sustainable forest management has continued to evolve through international forest policy dialogue, and has become rather multifarious and unspecific. The following seven thematic elements are now considered key components of sustainable forest management (FAO 2006):

(1) Extent of forest resources

(2) Biological diversity

(3) Forest health and vitality

(4) Productive functions of forest resources

(5) Protective functions of forest resources

(6) Socio-economic functions

(7) Legal, policy and institutional framework

These elements have to be considered when judging the sustainability of forest management. In this article on sustainable forest management in Ecuador it is not possible to take all seven elements into account. Emphasis will be put on the first two.

Sustainable forest management in Ecuador

Using forest area as a criterion for the evaluation of sustainable forest management in Ecuador, it is necessary to take account of the process of deforestation in the country. Assuming an original forest area of about 25 Mio hectares and a forest cover of 90% (Wunder 2000), the forest area in Ecuador is now reduced to 10.8 Mio hectares and forest cover to 39% (Table 1). About 7 Mio hectares of forest were lost in the last 50 years and forest area continues to decrease. The current deforestation rate in Ecuador is the highest in South America (-1.7%) with an annual loss of 198.000 hectares of forests (FAO 2006, Mosandl et al. 2008), which can largely be attributed to land use changes. Many forests have been converted into agricultural land and pastures. From 1972 to 1989 pasture area increased from 2.2 million hectares to 6 million hectares (Wunder 2000), which corresponds to 224,000 ha yr^{-1}. Assuming that the conversion of forests into pastures is continuing at the same rate, the current increase of agro-pastoral land equals the decrease in forest cover.

The loss of forests is an indication of non-sustainable forest management in Ecuador. Another sign of non-sustainability is the lack of reforestation efforts. The decrease of tropical mountain forests is not mitigated by an increase of forest plantations (Mosandl et al. 2008). No substantial areas were reforested in Ecuador during recent years. Indeed, plantation area increased only by 560 hectares per year during 2000-2005 (FAO 2006).

Table 1. Loss of forest area and forest cover in Ecuador.

year	forest area [10^6 ha]	forest area [%]	author
before human impact	25	90	Wunder 2000
1958	17.5	63	Cabarle et al. 1989
1987	12.5	45	FAO 1994
1990	11.9	43	FAO 1993
2005	10.8	39	FAO 2006

Options for sustainable forest management in tropical mountain forests of Ecuador

Concerns about the non-sustainability of management of the mountain forests in Ecuador were one reason for the establishment of DFG Research Unit 402. Research objectives of the unit included on the one hand the investigation of the high biodiversity in these mountain forests, on the other hand the development of management concepts aimed at the conservation of these forests. By describing the deforestation process in a selected area in southern Ecuador, it was hoped to gain insight in the way how to resolve the problem of forest destruction and biodiversity loss in the country. The process of deforestation was studied in the research area "Estación Científica San Francisco" (ECSF) by several research teams (Beck et al. 2008; see also chapters 4 and 13 of this volume). Mountain forests in this area are converted into pastures because pasture management seems to be more profitable than management of natural forests. In sites with high production potentials pasture management is again replaced by agriculture (home gardening), being more attractive in these sites. This undesirable process may be halted in two different ways, firstly by conservation of tropical mountain forests in protected areas, and secondly by ecologically and economically sustainable forest management. Correspondingly, human impact in the study area may be visualized by means of two different gradients (Figs. 1-2; Beck et al. 2008). In Fig. 1, land use classes are arranged along a "progressive" gradient of human disturbance, including (in order of decreasing naturalness): I) natural forests, II) managed forests, III) silvo-pastoral areas, IV) pastures devoid of trees, and V) areas of extensive agricultural use such as home gardens or crop fields. In Fig. 2, land use classes are arranged along a "regressive" gradient taking into account the processes of rehabilitation of deforested land, by reforestation (with native or exotic tree species), enrichment planting (with native species, following the planting of exotic ones) or natural succession. Classes include (in order of increasing naturalness): V) areas of extensive agricultural use, IV) abandoned pastures (often covered by bracken fern) left to succession, III) partially reforested areas for silvo-pastoral use, II) completely reforested areas (mixture of exotic and native trees), I) closed secondary forests with a high proportion of native species.

Naturalness in the research area is closely linked with high biodiversity. Arranging the different land use types according their biodiversity (as expressed by the Shannon index) exhibits a clear trend of increasing biodiversity with decreasing human impact (Table 2).

Table 2. Biodiversity in different land use types along the regressive gradient of human disturbance (in comparison with primary forest), as expressed by the Shannon Index. All types are located between 1800–2100 m a.s.l. Biodiversity increases with naturalness of land use type.

land use type	Shannon Index	author	measured compartments
pasture	0.9	Flick 2003	herbs, shrubs, trees
	ca. 1.6	Aguirre 2007	shrubs, trees
bracken fern	0.84	Flick 2003	herbs, shrubs, trees
	ca. 1.8	Aguirre 2007	shrubs, trees
shrubs	2.6	Flick 2003	herbs, shrubs, trees
	ca. 2.6	Aguirre 2007	shrubs, trees
secondary forest (38 yrs)	2.0	Günter 2007	woody plants (dbh > 2 cm)
pine plantation	2.3	Wirschmidt 2005	herbs, shrubs, trees
primary forest on ridges	4.61	Homeier et al. 2002	woody plants (dbh > 5 cm)
primary forest in valleys	4.91	Homeier et al. 2002	woody plants (dbh > 5 cm)

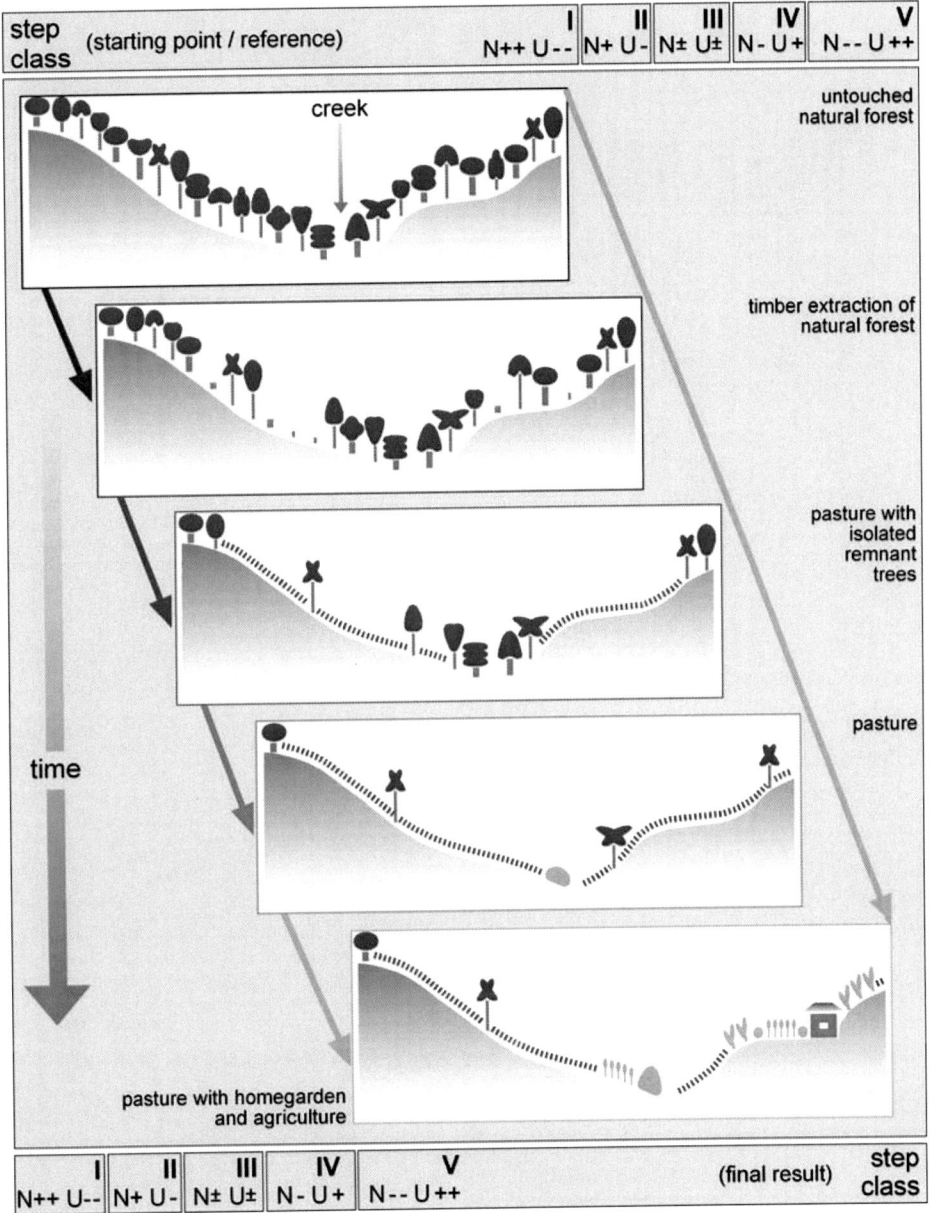

step class	(starting point / reference)	I N++ U--	II N+ U-	III N± U±	IV N- U+	V N-- U++

Figure 1. Progressive gradient (from ++ to − −) of human disturbance in the study area. Naturalness classes (N) and their intensity of use (U, steps I to V) are defined by their supposed distance to the reference areas "natural forest" and "extensive land use" (from Beck et al. 2008).

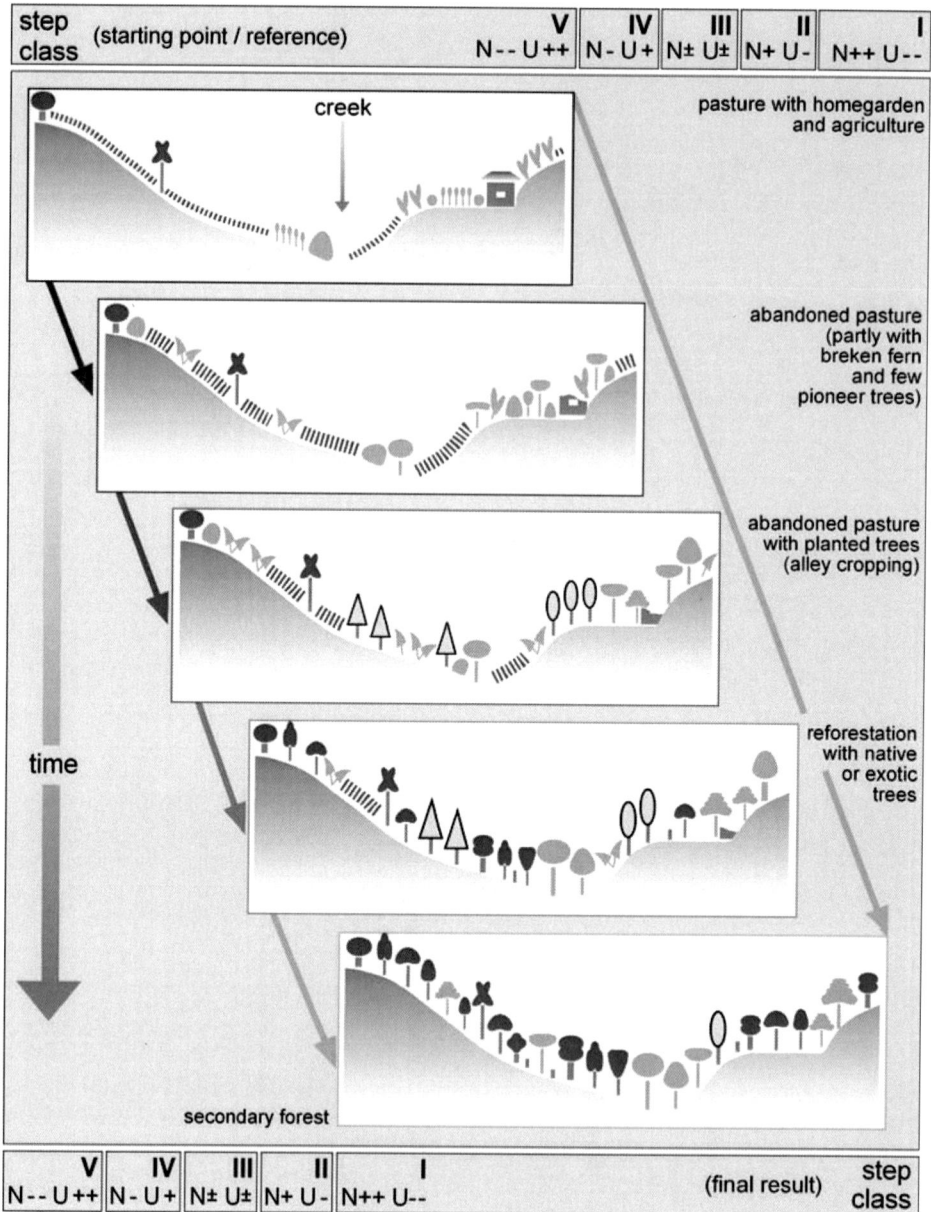

Figure 2. Regressive gradient (from ++ to – –) of human disturbance in the study area. Naturalness classes (N) are defined by their supposed distance to the reference areas "natural forest" and "extensive land use (U)" (from Beck et al. 2008).

To combat deforestation of the Ecuadorian mountain forests, concerted actions are necessary. On the one hand, *conservation* of forest areas and their inherent biodiversity is required as well as a scientifically-based *natural forest management*, to prevent conversion into pasture land. On the other hand, *succession* could in some cases enhance the return of agricultural land to a more natural state; subsequently, *reforestation* of degraded pastures and *enrichment planting* with native species might lead to secondary forests which can be used for timber production. All this will contribute to release the pressure from the pristine forests.

Within DFG Research Unit 402, first results towards the development of sustainable management of tropical mountain forest have been achieved. Some of these results are presented below.

Biodiversity conservation

Andean montane forests are characterized by high species richness (Brummit & Lughadha 2003, Henderson et al. 1991). The study area exhibits unusually high diversities of vascular plants (Homeier 2004, Homeier & Werner 2008), and especially of bryophytes (Gradstein et al. 2008) and geometrid moths (Brehm et al. 2005). Richter et al. (2008) attribute this high diversity to the complex interaction of various local biotic and abiotic factors. The extremely high climatic variability in the region (Richter 2003) leads to the existence of numerous genecological zones in the Province of Loja (Table 3). Günter et al. (2004) identified up to 134 potential genecological zones, 12.7% of these protected in National Parks, 50% protected outside national parks and 37% remaining without protection. These protected areas contribute greatly to the conservation of biodiversity, they help conserving valuable tree species in pristine forests and safeguarding their use in exploited ones.

Table 3. Area of potential genecological zones [km^2] under different climatic conditions in the province of Loja, South Ecuador (from Günter et al. 2004).

mean temperature [°C]	number of dry months					
	0 - 2	3 - 4	5 – 6	7 - 8	9 - 10	11 - 12
8 – 12	48	397	410	147	31	0
12 – 16	68	432	768	377	107	2
16 – 20	21	71	1165	1154	194	55
20 – 26	0	0	779	2349	2306	75
total	137	900	3121	4027	2637	132

Natural forest management

The extraordinary biodiversity of a region cannot only be conserved in national parks or protected areas. In addition to conservation, appropriate forest management is necessary to maintain biodiversity. In comparison to complete conversion of forest, sustainable forest management is considered to have low-intensity impact on biodiversity (Grau 2002) and is considered a suitable tool for biodiversity conservation as well as integration of forest conservation and economic development (Bawa & Seidler 1998, Chazdon 1998, Dawkins & Philip 1998). Sustainable management of highly diverse tropical forest ecosystems requires case-specific silvicultural approaches meeting the local socioeconomic and ecological requirements of management and conservation (Hutchinson 1993). Improvement felling, for example, is one of the oldest and widely applied silvicultural instruments (Dawkins & Philip 1998). Its feasibility for sustainable management of tropical mountain forests, however, remains largely unknown.

Table 4. Diameter increment of liberated and reference trees one year after liberation (from Günter et al. 2008).

species	liberated [mm/yr]	reference [mm/yr]	significances between species	timber value
Cedrela montana	13.3±1.9	18.05±2.3	bc*	A
Clusia cf. ducuoides	1.6±0.3	1.7±0.3	a*	C
Ficus citrifolia	17.8±8.0	20.8±4.5	d*	C
Hyeronima asperifolia	7.2±1.1	5.8±1.5	abc*	B
Hyeronima moritziana	2.4±0.5(*)	1.2±0.3(*)	a*	B
Inga acreana	14.9±4.3(*)	6.7±2.1(*)	c*	C
Nectandra membrana-cea	5.8±1.8	4.1±1.0	ab*	B
Podocarpus oleifolius	2.9±1.1	2.0±0.6	a*	A
Tabebuia chrysantha	2.5±0.8	3.8±0.7	a*	A

Significant differences at p = 0.1 indicated by (*), at p=0.05 by *. Significantly different growth rates between species indicated by letters (a, b, c, d). Timber values vary from high (A), medium (B) and low (C).

A selective logging experiment on 13 ha of undisturbed tropical mountain forest (ca. 1900 m a.s.l.) in the San Francisco Valley showed significant different growth rates between the studied tree species (Table 4). One year after the liberation five species (*Hyeronima asperifolia, Hyeronima moritziana, Inga acreana, Nectandra membranacea, Podocarpus oleifolius*) responded with an enhanced diameter increment, in others silvicultural treatments had slightly negative growth effects (*Tabebuia chrysantha*, *Cedrela montana*). It is expected that positive growth in the latter species may eventually occur in the subsequent years due to the reduced crown competition. *Cedrela montana* showed high growth rates and since it also represents a very high value timber species, it could be a key genus for natural forest management. Advantage should be taken of the high abundance of further valuable timber species despite of their rather lower growth rates, in order to increase the added value, especially *Ocotea* spp., *Nectandra* spp and other genera of Lauraceae.

Table 5. Basal area [m²/ha] of ca. 40 year old secondary montane forest (2100 m a.s.l.) at different distances to primary forest edge. Significant differences at the p = 0.05 level indicated by letters (from Günter et al. 2007).

dbh-class	primary forest	distance from the forest edge		
		0 m	20 m	40 m
> 2 cm	24.7 a	12.2 b	11.9 b	12.4 b
> 5 cm	21.2 a	8.1 b	7.0 b	7.9 b
> 10 cm	13,3	1,0	0,0	0,8

Natural succession

Reasons for slow natural re-colonization of deforested land may be remoteness from forest edges, reduced seed input and forest fragmentation (Cubiña & Aide 2001; Myster 2004). However, where distance to remaining forest edges is limited, natural regeneration might be a valuable option for the restoration of biodiversity. The regeneration of a ca. 40 y old secondary forest in the study area, developed on abandoned pasture land through natural succession, was analyzed in relation to distance from the surrounding primary forest (Günter et al. 2007). The results showed that regeneration had been rather slow, trees with dbh > 10 cm were still very scarce and basal area far below that of the primary forest (Table 5). Also, number of species at plot level and Shannon index were lower in the secondary forest as compared to the primary forest. Families which acted as generalists and were common in all primary forests, forest edges and secondary forests had higher absolute abundances than families which were found mainly in secondary and, especially, primary forests (Fig. 3). Furthermore, several primary forest families with high-value timber species (Podocarpaceae, Lauraceae, Bignoniaceae, Meliaceae) were absent in the regenerated forest. The results show that natural succession on abandoned pastures does not lead to forests with a high amount of valuable timber within a short period.

Reforestation

Reforestation may accelerate the restoration process when fast growing timber species are used, contributing to additional income. Most of the existing plantations in Ecuador consist of introduced species (*Pinus* spp., *Eucalyptus* spp.). Because of ecological problems with these species, there is a growing interest in the use of native species in reforestation (Brandbyge & Holm-Nielsen 1986; Aguirre et

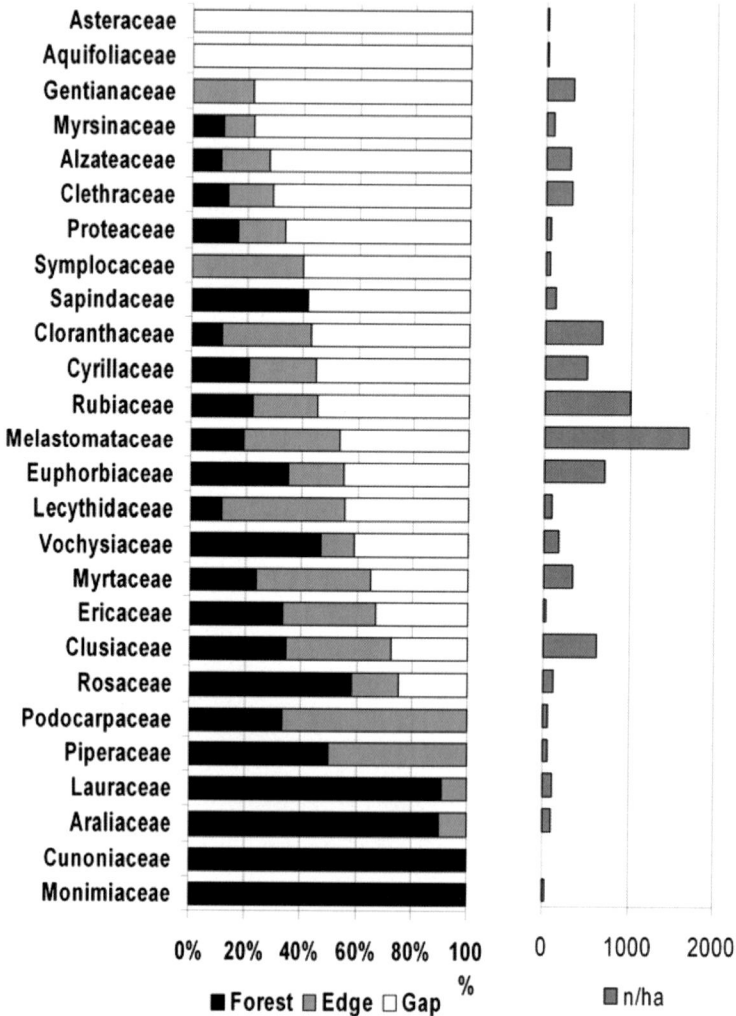

Figure 3. Mean relative abundances (left) and absolute abundances (right) of natural regeneration in a primary forest, forest edge and an adjacent secondary forest on a large gap (80 m x 240 m) created by pasture management ca. 40 years ago (from Günter et al. 2007).

al. 2002a, Aguirre et al. 2002b). However, the lack of genetically suitable seed and other reproductive plant materials has impeded the establishment of large scale reforestation with native species in Ecuador (Stimm et al. 2008).

Reforestation trials with five native and two exotic species (Fig. 4) on 12 ha in the study area revealed that *Alnus acuminata* can compete in height growth with the introduced species *Pinus patula* and *Eucalyptus saligna* in the first successional stages after pasture abandonment (active pasture management, bracken fern, shrub stage). As expected, the mid-successional and high-value timber species *Tabebuia chrysantha*, *Juglans neotropica* and *Cedrela montana* showed rather slow growth rates. The survival rate among the species differed greatly (Fig. 4). Survival of the two exotic ones was 92% but that of the native, relatively shade tolerant *T. chrysantha* was slightly higher (94%). Surprisingly, the survival of the light demanding *Alnus acuminata* and *Heliocarpus americanus* was significantly lower (57%) than that of all other species except *Juglans neotropica* (44%).

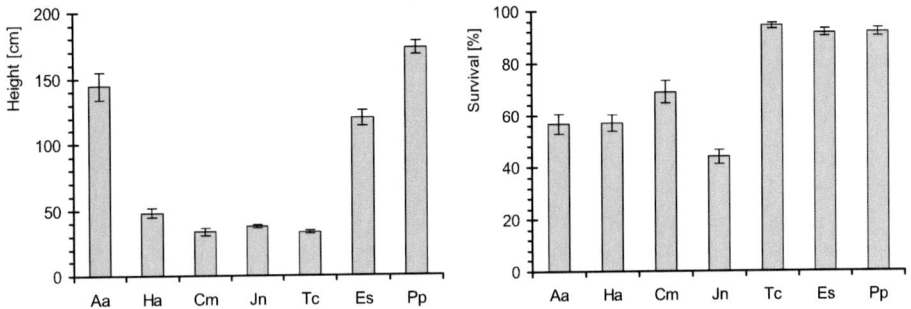

Figure 4. Survival and total height of tree species two years after planting (n=48 plots). Aa = *Alnus acuminata*, Cm = *Cedrela montana*, Es = *Eucalyptus saligna*, Ha = *Heliocarpus americanus*, Jn = *Juglans neotropica*, Pp = *Pinus patula*, Tc = *Tabebuia chrysantha* (from Aguirre 2007).

Enrichment planting

A further option for forest rehabilitation is enrichment planting under established natural succession. If the seed flux into abandoned areas is not sufficient, planting of animal-dispersed species like *Piper* spp. or *Myrica pubescens* may attract bats or birds (Almeida et al. 2004) and thus enhance the restoration processes. Enrichment planting with valuable timber species such as Podocarpaceae can provide additional income for small scale farmers. Enrichment planting trials in tropical mountain forest (e.g. Pedraza & Williams-Linera 2003) confirm that enrichment planting can accelerate natural succession, especially when animal-dispersed species are used (Garzia-Martinez & Howe 2003).

A further option is enrichment planting on highly degraded land. In the latter case two steps are necessary. First, exotic species are used to help restoring forest cover within a reasonable time span (Sabogal, 2005). Despite often-mentioned objections against the use of exotic taxa, there is evidence that they may help preventing habitat degradation processes through soil stabilization, increase of organic matter levels and improvement of soil nutritional status (Montagnini & Sancho 1990, Parrotta 1992, 1993, Lugo 1997, Feyera et al. 2002a, Carpenter et al. 2004). In a second step the established plantations can be used as a shelter for the enrichment with native species.

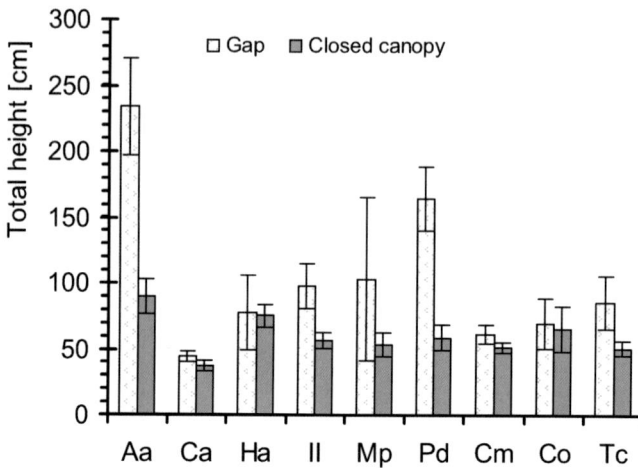

Figure 5. Height of nine native tree species two years after planting under forest canopy and in gaps (shown are means and standard errors). Aa = *Alnus acuminata*, Ca = *Cupania* cf. *americana*, Cm = *Cedrela montana*, Co = *Cinchona officinalis*, Ha = *Heliocarpus americanus*, Il = *Isertia laevis*, Mp = *Myrica pubescens*, Pd = *Piptocoma discolor*, Tc = *Tabebuia chrysantha* (from Aguirre 2007).

In our study, nine native species were planted in an established 20 year old pine plantation adjacent to the reforestation area mentioned above, under the forest canopy and in natural gaps. The results showed that all species grew better in the gaps (Fig. 5). Especially *Alnus acuminata* and *Piptocoma discolor* showed excellent growth and reached mean heights of 234 cm and 165 cm respectively in two years, being more than 2.5 times that of trees planted under the canopy and 1.6 times of those in the reforestation experiment. The valuable timber species *Cedrela montana* and *Tabebuia chrysantha* also achieved faster growth (Aguirre et al. 2006, Aguirre 2007). It was concluded that the establishment of plantations with exotic species and the subsequent enrichment with native species might be a promising option for the reforestation of abandoned pastures and rehabilitation of biodiversity.

Conclusions

As compared with the principles of sustainable forest management, forest management in Ecuador with the highest deforestation rate in South America is far from being sustainable. There are different strategies to overcome this problem. On the one hand, conversion of natural forests into pastures has to be decelerated. On the other hand, the rehabilitation of degraded land should be accelerated. Our research in tropical mountain forest of Ecuador has led to the following conclusions:

(1) Conservation of tropical mountain forests is a powerful means to secure valuable tree species and biodiversity.

(2) Natural forest management seems possible even though the effectiveness of improvement fellings has not yet been proven.

(3) Succession on abandoned pastures does not lead to forests with a high amount of valuable timber within a short period (40 y). Therefore, reforestation is necessary.

(4) Reforestation of degraded land (pastures) is possible with native as well with exotic tree species.

(5) Exotic tree species may act as nurse trees for the establishment of native tree species.

Successful sustainable forest management in Ecuadorian tropical mountain forests requires efforts in all these fields and a synopsis of the different strategies.

Acknowledgements. The authors gratefully thank the German Research Foundation (DFG) for funding the silvicultural project within the research groups FOR 402 and FOR 816, Nature and Culture International for logistic support, and the Ministerio del Ambiente, Ecuador, for research permits.

References

Aguirre N, Ordoñez L, Hofstede R (2002a) Identificación de Fuentes Semilleras de Especies Forestales Alto-Andinas. PROFAFOR, el Ministerio del Ambiente y ECOPAR, Reporte Forestal 6

Aguirre N, Ordoñez L, Hofstede R (2002b) Comportamiento Inicial de 18 Especies Forestales Plantadas en el Páramo. PROFAFOR, el Ministerio del Ambiente y ECOPAR, Reporte Forestal 7

Aguirre N, Günter S, Weber M, Stimm B (2006) Enriquecimiento de plantaciones de Pinus patula con especies nativas en el sur del Ecuador. Lyonia 10: 17-29

Aguirre N (2007) Silvicultural contributions to the reforestation with native species in the tropical mountain rain forest region of South Ecuador. Dissertation, Institute of Silviculture, Technical University München

Almeida K, Arguero A, Clavijo X, Matt F, Zamora J (2004) Dispersion de semillas por aves, murcielago y viento en áreas disturbadas de un bosque montano en el suroriente ecuatoriano. Annual Symposium, DFG Research Unit 402, Loja

Bawa KS, Seidler R (1998) Natural forest management and conservation of biodiversity in tropical forests. Conservation Biology 12: 46-55

Beck E Mosandl R, Richter M, Kottke I (2008) The investigated gradients. In: Beck E, Bendix J, Kottke I, Makeschin F, Mosandl R (eds) Gradients in a Tropical Mountain Ecosystem of Ecuador. Ecological Studies 198. Springer, Berlin, Heidelberg, New York, pp 55-62

Brandbyge J, Holm-Nielsen LB (1986) Reforestation of the high Andes with local species. Reports from the Botanical Institute, University of Aarhus 13

Brehm G, Pitkin LM, Hilt N, Fiedler K (2005) Montane Andean rain forests are a global diversity hotspot of geometrid moths. Journal of Biogeography 32: 1621-1627

Brummit N, Lughadha EI (2003) Biodiversity. Where's Hot and Where's Not. Conservation Biology 17: 1442-1448

Cabarle BJ, Crespi M, Dodson CH, Luzuriaga C, Rose D, Shores JN (1989) An assessment of biological diversity and tropical forests for Ecuador. USAID, Quito

Carlowitz HC von (1713) Sylvicultura oeconomica. Anweisung zur wilden Baum-Zucht. Reprint. Bibliothek "Georgius Agricola" 135, Bergakademie Freiberg, Freiberg

Carpenter F, Nicols J, Pratt T, Young K (2004) Methods of facilitating reforestation of tropical degraded land with the native timber tree *Terminalia amazonica*. Forest Ecology and Management 202: 281-291

Chazdon RL (1998) Tropical Forests—Log ´em or leave ´em? Science 281: 1295-1296

Cubiña A, Aide TM (2001) The effect of distance from forest edge on seed rain and soil seed bank in a tropical pasture. Biotropica 33: 260–267

Dawkins HC, Philip MS (1998) Tropical moist forest silviculture and management. CAB International, United Kingdom

FAO (1993) Forest Resources Assessment 1990: Tropical countries. FAO, Rome

FAO (1994) Forest Resources Assessment 1990: Country briefs. FAO, Rome

FAO (1999) State of the world's forests 1999. FAO, Rome

FAO (2001) State of the world's forests 2001. FAO, Rome

FAO (2006) Global Forest Resources Assessment 2005. Progress towards sustainable forest management. FAO, Rome

Feyera S, Beck, E, Lüttge U (2002) Exotic trees as nurse-trees for the regeneration of natural tropical forests. Trees 16: 245-249

Fimbel R, Fimbel C, (1996) The role of exotic conifer plantation in rehabilitating degraded tropical forest land: a case study the Kibale forest in Uganda. Forest Ecology and Management 81: 215-226

Flick C (2003) Vegetationskundliche Untersuchungen zu kleinstandörtlichen Unterschieden aufgelassener Weideflächen verschiedener Entwicklungsstufen in Südecuador. Diploma thesis, Institute of Silviculture, Technical University München

Garzia-Martinez C, Howe HF (2003) Restoring tropical diversity: beating the time tax on species loss. Journal of Applied Ecology 49: 423–429

Gradstein SR, Bock C, Mandl N, Nöske N (2008) Bryophyta. In: Liede-Schumann S, Breckle W (eds) Provisional checklists of fauna and flora of the San Francisco valley and its surroundings (Reserva San Francisco/Prov. Zamore–Chinchipe, southern Ecuador). Ecotropical Monographs 4: 69-87

Grau, HR (2002) Scale dependent relationships between treefalls and species richness in a neotropical montane forest. Ecology 83: 2591-2601

Grober U (2000) Der Erfinder der Nachhaltigkeit – Hans Carl Edler von Carlowitz. Einleitung in: Carlowitz, H.C. von (1713) Sylvicultura oeconomica. Anweisung zur wilden Baum-Zucht. Reprint. Bibliothek "Georgius Agricola" 135, Bergakademie Freiberg, Freiberg

Günter S, Stimm B, Weber M (2004) Silvicultural contributions towards sustainable management and conservation of forest genetic resources in Southern Ecuador. Lyonia 6: 75-91

Günter S, Weber M, Erreis R, Aguirre N (2007) Influence of forest edges on the regeneration of abandoned pastures in the tropical mountain rain forest of Southern Ecuador. European Journal of Forestry Research 126: 67-75

Günter S, Mosandl R, Cabrera O, Zimmermann M, Fiedler K, Knuth J, Boy J, Wilcke W, Meusel S, Makeschin M, Werner F, Gradstein SR (2008) Natural Forest Management in Neotropical Mountain Rain Forests: An Ecological Experiment. In: Beck E, Bendix J, Kottke I, Makeschin F, Mosandl R (eds) Gradients in a Tropical Mountain Ecosystem of Ecuador. Ecological Studies 198. Springer, Berlin, Heidelberg, New York, pp 363-375

Hasel K (1985) Forstgeschichte. Ein Grundriß für Studium und Praxis. Pareys Studientexte 48. Paul Parey, Hamburg, Berlin

Henderson A, Churchill SP, Luteyn JL (1991) Neotropical plant diversity. Nature 351: 21-22

Homeier J, Dalitz H, Breckle S-W (2002) Waldstruktur und Baumartendiversität im montanen Regenwald der Estación Científica San Francisco in Südecuador. Berichte der Reinhold-Tüxen-Gesellschaft 14: 109-118.

Homeier J (2004) Baumdiversität, Waldstruktur und Wachstumsdynamik zweier tropischer Bergregenwälder in Ecuador und Costa Rica. Dissertationes Botanicae 391: 1-207

Homeier J, Werner F (2008) Spermatophyta. In: Liede-Schumann S, Breckle W (eds) Provisional checklists of fauna and flora of the San Francisco valley and its surroundings (Reserva San Francisco/Prov. Zamore–Chinchipe, southern Ecuador). Ecotropical Monographs 4

Howe HF, Smallwood J (1982) Ecology of seed dispersal. Annual Review of Ecology and Systematics 13: 201–228

Hutchinson, ID (1993) Puntos de partida y muestreo silvicultural de bosques naturales del trópico húmedo. Colección, silvicultura y manejo de bosques naturales No 7. Informe Técnico No 204, CATIE, Turrialba

Lugo AE (1997) The apparent paradox of reestablishing species richness on degraded lands with tree monocultures. Forest Ecology and Management 99: 9-19

Montagnini, F., Sancho, F., 1990. Impacts of native trees on tropical soils: a study in the Atlantic lowlands of Costa Rica. Ambio 19: 386-390

Mosandl R, Günter S, Stimm B, Weber M (2008) Ecuador suffers the highest deforestation rate in South America. In: Beck E, Bendix J, Kottke I, Makeschin F, Mosandl R (eds) Gradients in a Tropical Mountain Ecosystem of Ecuador. Ecological Studies 198. Springer, Berlin, Heidelberg, New York, pp 37-40

Myster W (2004) Regeneration filters in post-agricultural fields in Puerto Rico and Ecuador. Vegetatio 172: 199–209

Parrotta JA, Turnbull J, Jones N, (1997) Catalyzing native forest regeneration on degraded tropical lands. Forest Ecology and Management 99: 1-7

Pedraza RA, Williams-Linera G (2003) Evaluation of native tree species for the rehabilitation of deforested areas in a Mexican cloud forest. New Forestry 26: 83–99

Richter M (2008) Vegetation structures and ecological features of the upper tree line ecotone In: Beck E, Bendix J, Kottke I, Makeschin F, Mosandl R (eds) Gradients in a Tropical Mountain Ecosystem of Ecuador. Ecological Studies 198. Springer, Berlin, Heidelberg, New York, pp 123-136

Sabogal C (2005) Site-level rehabilitation strategies for degraded forest lands. In: International Tropical Timber Organization (ITTO). Restoring forest landscapes: an introduction to the art and science of forest landscape restoration. ITTO, Technical Series 23: 101-108

Schanz H (1996) Forstliche Nachhaltigkeit – Sozialwissenschaftliche Analyse der Begriffsinhalte und -funktionen. Institut für Forstökonomie, Universität Freiburg

Stimm B, Beck E, Günter S, Aguirre N, Cueva E, Mosandl R, Weber M (2008) Reforestation of abandoned pastures: Seed ecology of native species and production of indigenous plant material. In: Beck E, Bendix J, Kottke I, Makeschin F, Mosandl R (eds) Gradients in a Tropical Mountain Ecosystem of Ecuador. Ecological Studies 198. Springer, Berlin, Heidelberg, New York, pp 433-446

Weber-Blaschke G, Mosandl R, Faulstich M (2005) History and mandate of sustainability: From local forestry to global policy. In: Wilderer PA, Schroeder ED, Horst Kopp H (eds) Global Sustainability. The Impact of Local Cultures. Wiley, Weinheim, pp 5-19

Wirschmidt B (2005) Vegetationskundliche Untersuchungen auf aufgelassenen Weideflächen und in Pinus patula Beständen in Südecuador. Diploma thesis, Institute of Silviculture, Technical University München

Wunder S (2000) The economics of deforestation: the example of Ecuador. St. Martin's press, New York

Biodiversity and Ecology Series (2008) 2: 195-217
The Tropical Mountain Forest – Patterns and Processes in a Biodiversity Hotspot
edited by S.R. Gradstein, J. Homeier and D. Gansert
Göttingen Centre for Biodiversity and Ecology

Ecological aspects of a biodiversity hotspot in the Andes of southern Ecuador

Erwin Beck[1] and Michael Richter[2]

[1]Department of Plant Physiology, University of Bayreuth, Universitätsstr. 30, 95440 Bayreuth, Germany, erwin.beck@uni-bayreuth.de
[2]Institute of Geography, University of Erlangen, Kochstr. 4/4, 91054 Erlangen, Germany

Abstract. Results and conclusions obtained by several research groups in the tropical mountain rainforest of the Reserva Biológica San Francisco, southern Ecuador, are used to reflect upon this exceptional hotspot of biodiversity from an ecosystem point of view. Features of functionality are emphasized while addressing the following questions: (1) What is a "biodiversity hotspot"? Can a relatively small area like the Reserva Biológica San Francisco be considered a hotspot of biodiversity? Does the term hotspot pertain to all groups of organisms of the area? How can we interprete species turnover along gradients? (2) Which geographical and ecological factors may have contributed to the accruement of the extraordinary organismic diversity of the area? (3) Which factors may be important for the maintenance of the hotspot? Which role does disturbance play? (4) Is there a feedback of biodiversity on ecosystem processes? (5) Is there redundancy of species in the South Ecuadorian hotspot?

Introduction

This chapter reflects upon the incidence of biodiversity hotspots from an ecosystem perspective. Biodiversity on a global scale has been censused for vascular plants (Barthlott et al. 2005) and based on this, centres of plant biodiversity with more than 5000 species of vascular plants per 10,000 km² have been identified. One of these centres, commonly called "biodiversity hotspots", is the tropical Andes-Amazonia region. It comprises dry and humid mountain areas up to more than 6000 m a.s.l. as well as lowland rain forests, and thus a great variety of ecosystems. These areas have not been examined in detail with respect to underlying, ecologically significant factors.

The first question arising is whether a specific ecosystem, or an ecologically explorable part of that ecosystem, may be considered representative of the biodiversity hotspot. We will present a "bottom-up" approach to deal with this question.

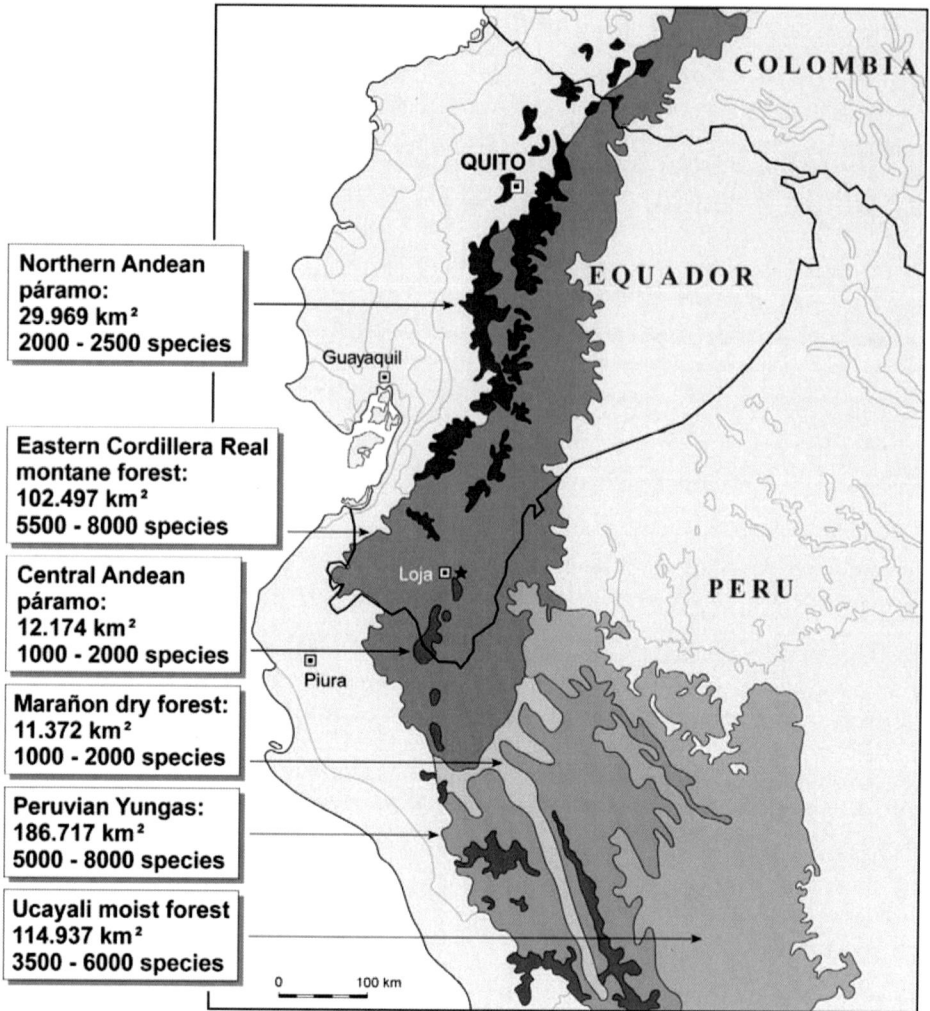

Figure 1. Ecoregions in the Neotropics (from Kier et al. 2005, based on Morrison et al. 2001). The black star indicates the location of the Reserva Biológica San Francisco.

On a global scale, distribution of terrestrial biodiversity has been systematically dealt with on the basis of plant species diversity. Ecosystem aspects, however, address interactions, e.g. of plants and animals. The question, therefore, is posed whether the term "hotspot" pertains to all groups of organisms of that area. When considering a biodiversity hotspot under environmental aspects, geographical, ecological and historical factors which may have contributed to the accruement of that extraordinary diversity are of special interest, as are factors important for the maintenance of the hotspot. Finally, it is tempting to examine the influence of biodiversity on ecosystem processes and explore whether, under the aspect of ecosystem functionality, redundancy of species can be assessed.

In the following these questions will be discussed using the results of a unique, comprehensive study of a tropical mountain rainforest in the southern Ecuadorian Andes. This study has been carried out by an interdisciplinary group of researchers over a period of ten years. The results of the investigations have been synthesized in a recent volume of "Ecological Studies" (Beck et al. 2008) and in a comprehensive species checklist (Liede-Schumann & Breckle 2008). A brief description of the research area, the "Reserva Biológica San Francisco" (RBSF), is given below.

Is the RBSF representative of the biodiversity hotspot "Tropical Andes of Ecuador"?

The studies of the research group in southern Ecuador covered an area of maximally 100 km², including a core area of 11 km² (the Reserva Biológica San Francisco, RBSF) and some satellite areas. The question arises whether this area is representative of the Andes-Amazonia hotspot and whether it can be addressed as a hotspot of biodiversity in a more general sense. For this purpose, a scale-down approach aimed at estimating the degree of biodiversity of the RBSF was performed, making use of the "ecoregion" concept. Ecoregions (Fig. 1) are relatively large units of land delineated to reflect boundaries of natural communities of animal and plant species. Using plant species richness data of 1800 "operational geographical units", 867 terrestrial ecoregions have been recognized on a world-wide basis (Kier et al. 2005). Highest richness is in the Borneo lowlands ecoregion (10,000 species), followed by nine ecoregions in Central and South America with ≥ 8000 species each. All of these are within the tropical and subtropical moist broadleaf forests biome. To compare the richness of plant species in a geographical region with that of an ecoregion, the number of species and the size of the respective areas are related using the following equation:

$$S_e = S_u \left(\frac{A_e}{A_u} \right)^z$$

where S_e = number of species in the ecoregion (here 8000); S_u = number of species in the geographical unit (see Table 1); A_e, A_u = area sizes (102,000 km² for Ecoregion NT 0121; 100 km² for the extended RBSF, respectively); z = a richness parameter (0.32 for tropical forests of South America) (Kier et al. 2005).

As shown in Table 1, more than 1450 species of vascular plants have been identified in the RBSF. According to the above equation, only 900 species would be required to meet the number of the ecoregion. In spite of the presumably better knowledge of the plant diversity of the RBSF, this relatively small area turned out an outstanding hotspot of plant diversity. Although ecoregion data are not available for cryptogams, the total of more than 500 species of bryophytes (mosses, liverworts, hornworts; Table 1) recorded from the RBSF may also be considered outstanding.

Table 1. Species diversity in the RBSF area and the San Francisco Valley, respectively. * RBSF only; *(x)* San Francisco valley, 1000-3000 m; ** world record.

	families	species	reference
seed plants	131	1208	Homeier & Werner (2008)
ferns and ferns allies	22*	250*	Lehnert et al. (2008)
hornworts	2*	3*	Gradstein et al. (2008)
liverworts	30*	320*	Gradstein et al. (2008)
mosses	40	204	Kürschner et al. (2008)
lichens	46*	311*	Nöske et al. (2008)
Glomeromycota		83*	Haug et al. (2004), Kottke et al. (2008)
Ascomycota		4*	Haug et al. (2004)
Basidiomycota		96* *(102)*	Haug et al. (2004)
bats		21* *(24)*	Matt (2001)
birds		227* *(379)*	Paulsch (2008), Rasmussen et al. (1994)
moths		*(2396)***	Brehm et al. (2005), Hilt (2005), Brehm et al. (2003), Fiedler et al. (2008)
butterflies (Papilionidae)		*(243)*	Häuser et al. (2008)
Bush crickets		*(101)*	Braun (2002)
soil mites (Oribatidae)		129* *(167)*	Illig & Maraun (2008)
soil amoebae (Testacea)		78* *(110)*	Krashevskay et al. (2008)

Only few other groups of organisms of the RBSF have been investigated in similar detail as plants. An extraordinary species-rich group are moths, for which a world record of species diversity has been recorded (Table 1; Brehm et al. 2005). With respect to vertebrates, only birds and bats have thus far been investigated; these groups likewise show outstanding diversity. Fungi have not been systematically surveyed and the species numbers given in Table 1 are by no means exhaustive (I. Kottke, pers. com.).

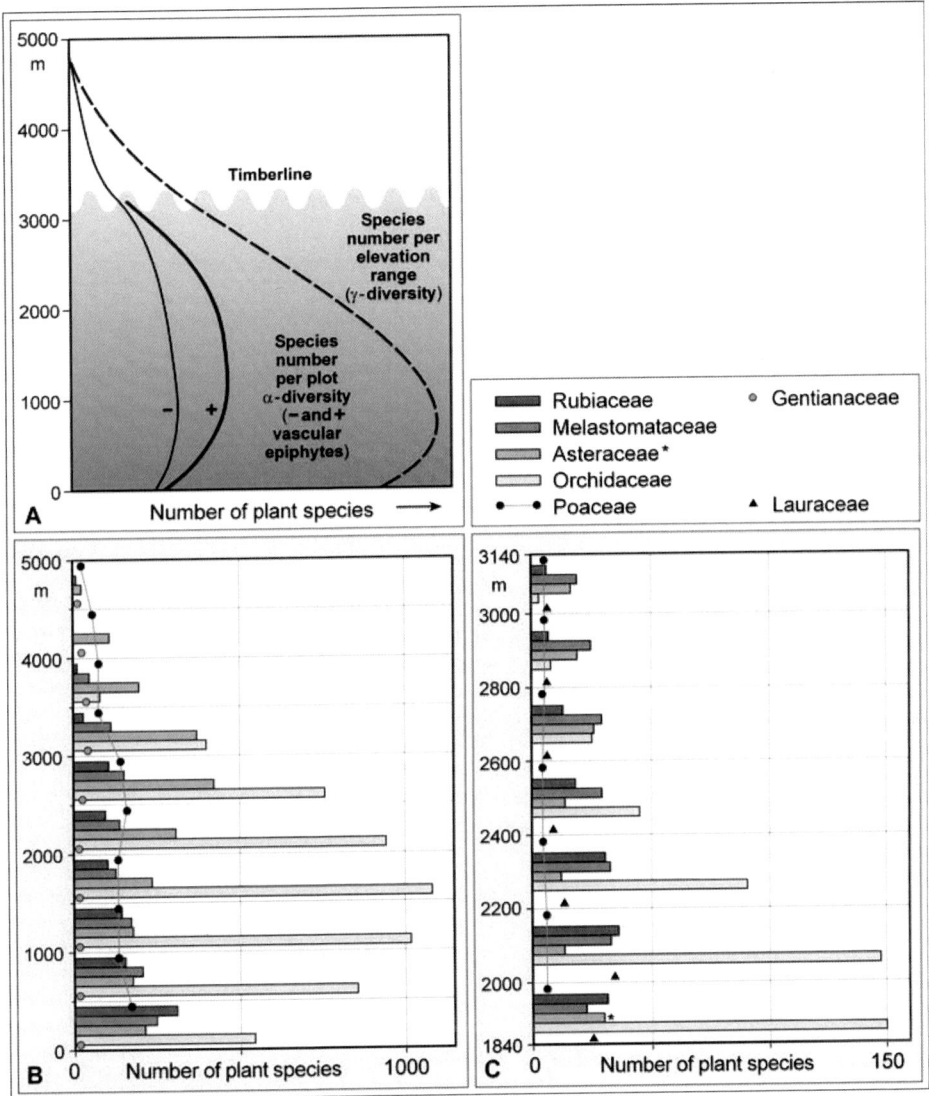

Figure 2. Rapoport's Rule along the elevational gradient and the Mid-Domain-Effect. **A,** altitudinal change of vascular plant α- and γ-diversities in tropical mountains (from Richter 2001, modified); species diversity and altitude are inversely correlated but there is usually a peak of richness at medium elevations (Mid-Domain-Effect). **B,** discordant vertical distribution of species numbers of selected plant families in Ecuador (data from Jørgensen & León-Yánez 1999). **C,** same as B, but for the RBSF area (data from Homeier & Werner 2008, supplemented by recent findings of K. H. Diertl, pers. com.).

Other groups of organisms such as litter decomposers, however, are relatively poor in species. Only a few earthworms have been found (M. Maraun, pers. com.), and soil mites (*Oribatidae*) and soil amoebae (*Testacea*), which have been carefully investigated, are disappointingly species-poor (Illig et al. 2005, Maraun et al. 2008; Table 1; see also chapter 6 of this volume). The available data, thus, show that the investigated area in spite of its small size can be considered to represent a hotspot of neotropical biodiversity, but not for all groups of organisms.

Why is it hot?

To explain high biodiversity, one has to differentiate between factors that have contributed to the development of this diversity and those that have contributed to its maintenance. Both types of factors are not always separable because those effective in maintaining biodiversity may also have been effective its accruement.

Positional effects in general: Rapoport's rule and the mid-domain effect. Rapoport's rule as exemplified by Stevens (1989, 1992, 1996) states that latitudinal ranges of plant and animal species are generally smaller at low than at high latitudes. Narrow ranges, as are typical of the Tropics, facilitate more species to coexist and thus result in higher species richness. It is evident that Rapoport's rule is subject to modification by factors such as drought or elevation.

The tropical regions also experienced a longer evolutionary time period since the Tertiary as compared with temperate regions and, thus, an elevated radiative speciation can be expected. Due to mostly favourable conditions for plant and animal life, newcomers or new ecotypes have better chances and thus the percentage of accidentals is comparatively high. Rapoport's rule may also apply to altitudinal gradients, posing that species ranges increase with altitude concomitantly with a decrease of species richness (Fig. 2). However, many studies on the altitudinal dimension of species diversity have shown that there is usually a peak of richness at medium elevations, the so-called Mid-Domain-Effect (Herzog et al. 2005, Krömer et al. 2005, Rahbek 2005; Fig. 2A; see also chapter 3 of this volume). Expectedly (taking into account the different ecological demands of certain plant families), the altitudinal distribution of various plant families is discordant, i.e. their mid-elevation peaks occur at different altitudes (Fig. 2B). The RBSF, at only some 450 kilometres south of the equator, occupies a latitudinal as well as an altitudinal range (1000–3200 m) where species richness can be assumed to be at its highest, as will be discussed below.

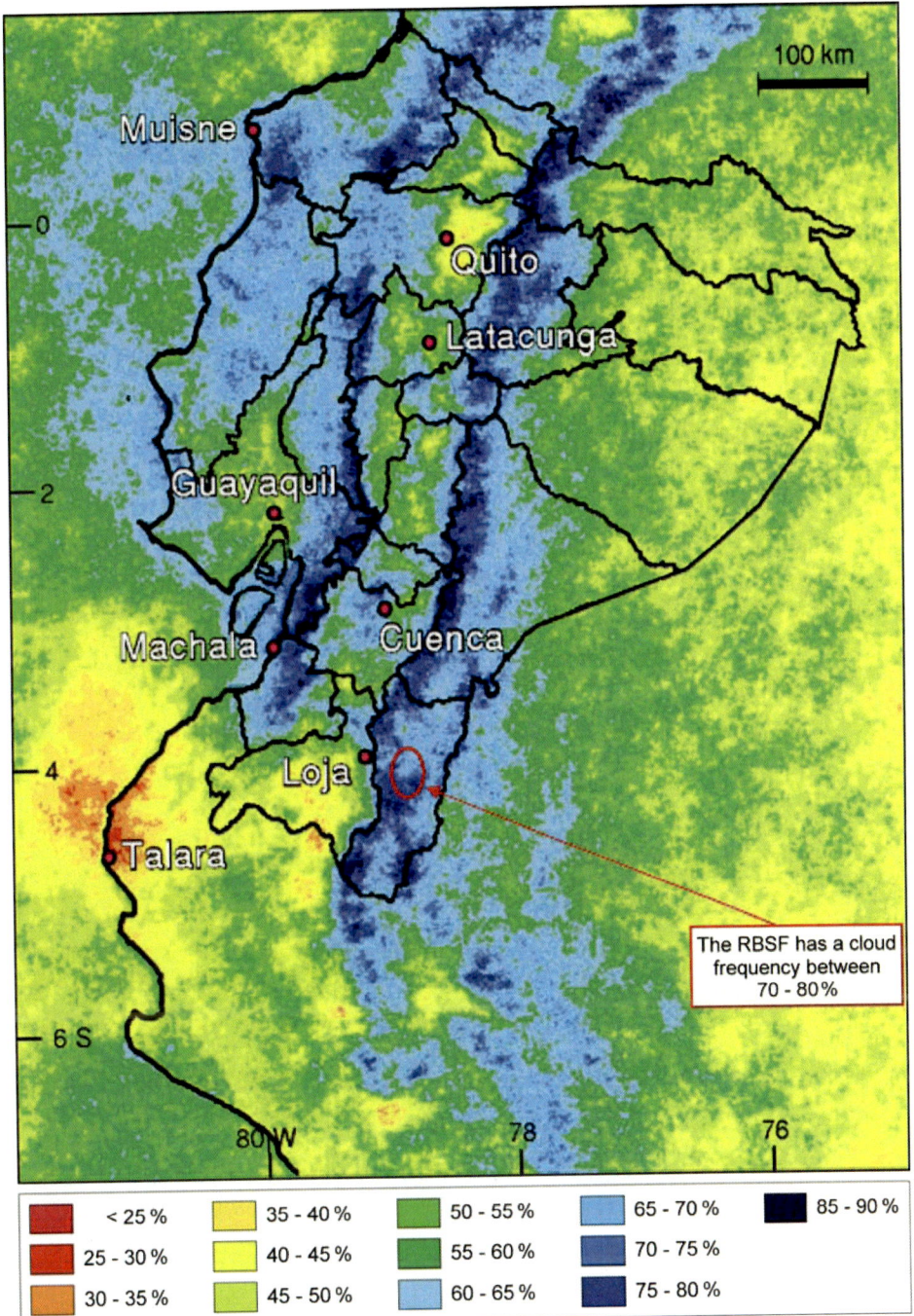

100 km

The RBSF has a cloud
frequency between
70 - 80%

< 25 %	35 - 40 %	50 - 55 %	65 - 70 %	85 - 90 %
25 - 30 %	40 - 45 %	55 - 60 %	70 - 75 %	
30 - 35 %	45 - 50 %	60 - 65 %	75 - 80 %	

Figure 3. Relative cloud frequency for Ecuador (2002-2003) and adjacent areas derived from NOAA-AVHRR data. From Bendix et al. (2004).

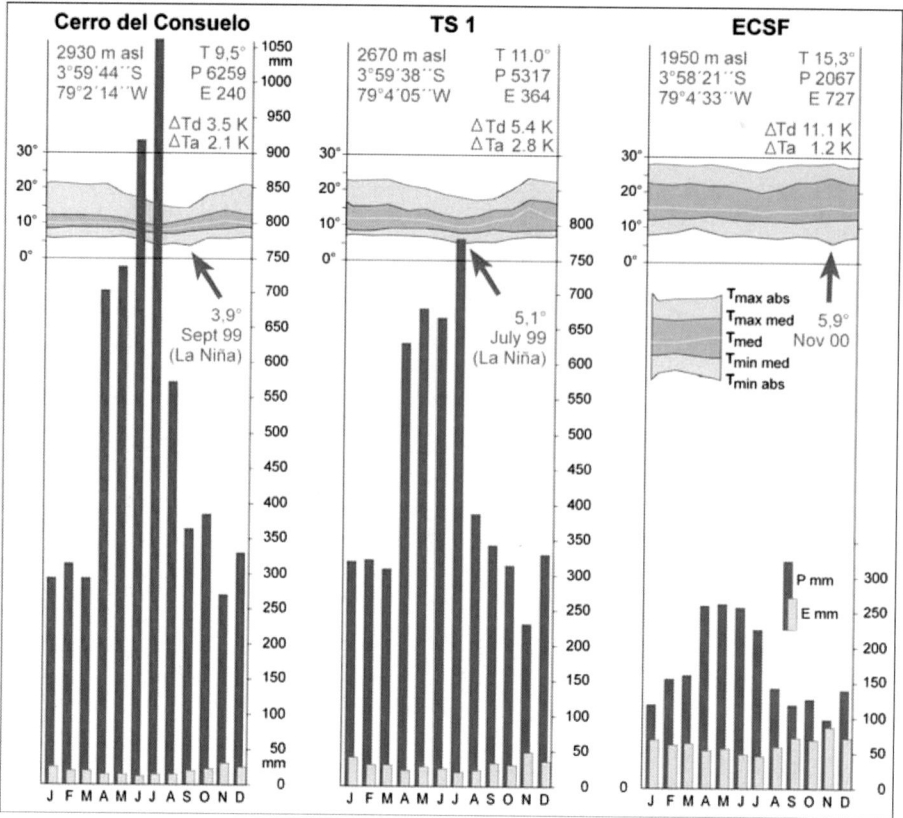

Figure 4. Climate diagrams of the RBSF showing the altitudinal gradients (monthly means) of precipitation, evaporation, amplitude of air temperature as well as the annual means. Temperature minima are highlighted by arrows. From Bendix et al. (2006), modified.

Positional effect of the RBSF. Located in the eastern range (= Cordillera Real) of the South Ecuadorian Andes, the RBSF belongs to the zonobiome of the humid tropics. The reserve is in the Amotape-Huancabamba depression, stretching from southern Ecuador to northern Peru, where the Andes barely reach 4000 m and the treeline is as low as ca. 3000 m. The geological setting of the Cordillera Real is rather monotonous and belongs to the paleozoic Chiquinda unit, consisting of metasiltstones, sandstones and quartzites, interspersed with layers of phyllite and clay (Litherland et al. 1994; Beck et al. 2008b). However, the terrain of this depression is extremely broken, e.g. in the "Nudo de Loja" from where several mountain ranges are stretching south- and southeastward with a great variety of differently exposed slopes. This rugged landscape harbours an extraordinary multiplicity of geomorphological microsites. For most of the year the Cordillera Real is subjected to strong easterly winds carrying a lot of moisture from the Atlantic Ocean over the Amazon. When reaching the Andes this moisture condenses with the rising air masses, causing the eastern slopes and the crest to be more or less constantly covered by clouds. A similar situation exists on the western escarpment of the Andes, which is subjected to the westerly winds from the Pacific Ocean. However, the Inner-Andean basin in the rain shadow of both wind systems receives only a small amount of precipitation as is shown by the cloud frequency (Fig. 3). With respect to the area of the Cordillera Real and the RBSF, an extraordinarily steep gradient of annual precipitation occurs over a short distance of not more than 30 km (Fig. 4). Likewise, a steep altitudinal precipitation gradient has been demonstrated in the area, from 2400 mm at 1850 m to over 6000 mm at 3100 m (Richter et al. 2008; chapter 1 of this volume).

The multiplicity of morphological microsites in combination with the strong climatological gradients result in a great variety of habitat types, enhancing the development of a high biological diversity. The incidence of a biodiversity hotspot in the South Ecuadorian Andes is in line with a recent model of plant diversity based on climate-driven growth conditions (Kleidon & Mooney 2000).

Landscape history. During the Quaternary, the Andean area including the Amotape-Huancabamba depression was subjected to strong climate changes (Niemann & Behling 2007; chapter 2 of this volume). With respect to plant diversity, the alternating glaciations and interglacial dry periods must have affected the local spreading of plant species, especially from the East to the West and *vice versa*. Jørgenson et al. (1995) described putative pleistocenic migration barriers of plants in the Ecuadorian Andes and recognized four regions of endemism (Fig. 5). Thus, the conceivable idea of the Amotape-Huancabamba-depression as a corridor for species exchange between the Amazon and the Pacific area as well as between the northern and the southern part of the central Andes, is less likely. Nevertheless, it cannot be ruled out that climatic changes resulted in periods of contact and separation between these areas, relative to the absence or presence of migration barriers. The Amotape-Huancabamba-depression might thus have functioned as a meeting point for the extraordinary genetic variety of the lowlands and as a centre of explosive radiation, e.g. in the genera *Anthurium, Piper* and *Cavendishia* (Gentry & Dodson 1987). Accruement of gene flow barriers by habitat fragmentation likewise results in enhanced genetic diversification and fosters endemism. Reports of recent expeditions by botanists exploring the flora of the Cordillera del Condor, Cordillera Colán and the Andean rainforest refugia in northern Perú have yielded large numbers of new species, all from the so-called "Amotape-Huancabamba floristic zone" (Dillon et al. 1995, Sagástegui et al. 1999, 2003, Weigend 2002, 2004). On the plant family level, far above-average degrees of local endemism have been demonstrated for Orchidaceae (55% of the occurring species), Bromeliaceae (50%), Asteraceae (37%), Piperaceae (37%) and Solanaceae (percentage not known).

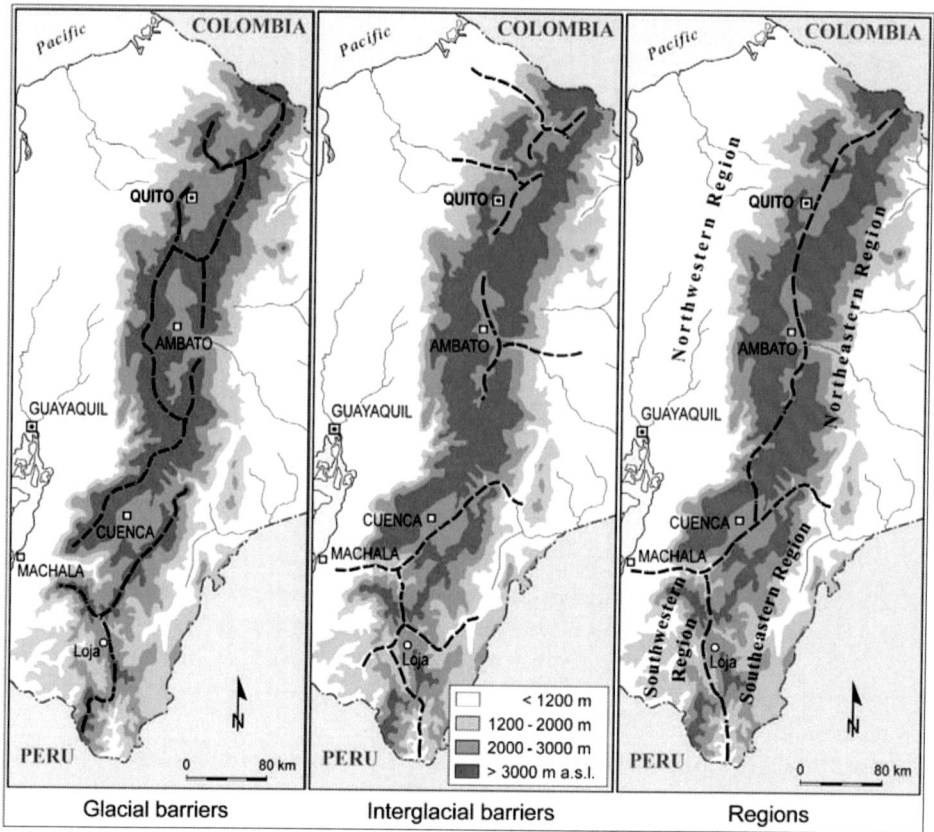

Figure 5. Possible migration barriers for plant species in the Ecuadorian Andes during the Pleistocene. *Left*: barriers due to Pleistocene glaciation. *Centre*: barriers due to dry land areas in large valleys. *Right*: recent situation: 4 regions of endemism. From Jørgensen et al. (1995), slightly modified.

An impressive account of the effects of climate change on the vegetation of the southern Ecuadorian Andes during the Holocene has been given by Niemann & Behling (2007; see also chapter 2 of this volume). Based on pollen analysis data these authors showed a change of the páramo vegetation in the San Francisco valley area from a grass-dominated herbaceous formation to the current shrubby subpáramo dominated by bushes and small trees. Climate warming after the disappearance of the glaciers caused an uplift of the treeline by presumably several hundred meters. Nevertheless, the recent treeline around 3000 m a.s.l. on the mountains of the Amotape-Huancabamba-depression is about 1000 to 1500 m lower than the treeline North and South of the depression. The comparatively low elevation of the treeline can be attributed to the continuous strong easterly winds and the extremely high precipitation in the eastern ranges, resulting in large areas of water-logged soil which prevent the growth of deep-rooting trees. This explanation is corroborated by two observations: (1) The treeline in the depression is composed of a great variety of woody plants, on average 70 species (Richter et al. 2008), whereas beyond the depression the treeline is formed only by *Polylepis* spp. (Rosaceae) occurring up to 4500 m and even higher; (2) Species-rich islets of small trees can be found in local, wind-sheltered hollows well above the treeline in the depression, indicating wind as an important factor inhibiting the growth of the taller plants (Fig. 6).

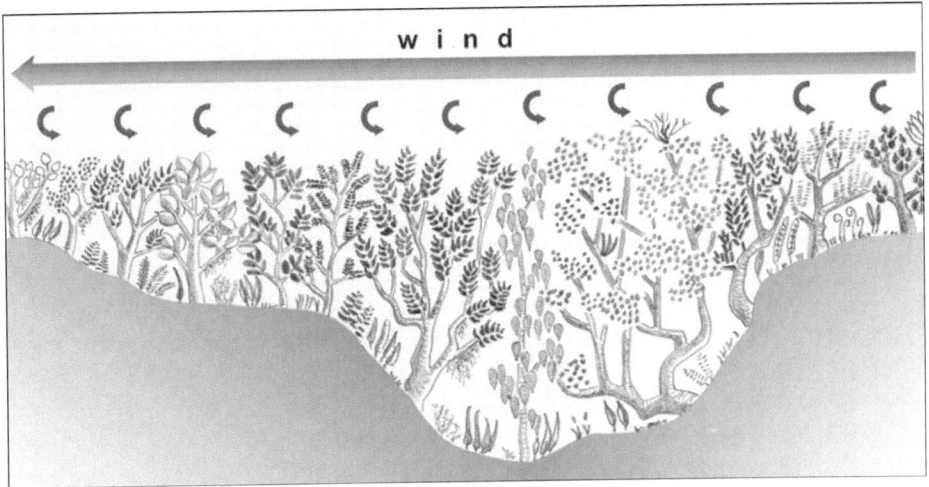

w i n d

Figure 6. The treeline in the Amotape-Huancabamba depression is subjected to continuously strong easterly winds forcing treelets into hollows and ravines, with only dwarf shrubs thriving on slopes and crests. From Richter (2001), modified.

Vegetation dynamics. In contrast to the very effective long-term climatic changes, short-term changes apparently are of lesser significance. In the eastern mountains of South Ecuador, the recurrent El Niño and La Niña periods result in an transitory decrease of precipitation (Holmgren et al. 2001), which may enhance plant growth in the area (Cueva et al. 2006). However, there are other dynamic phenomena which could have an effect on plant diversity. First, tropical forests like those of the RBSF exhibit a so-called mosaic-climax characterized by a pronounced patchiness of different successional stages. The simultaneous coexistence of early and late pioneers and species characteristic of a mature forest enhances an increased plant diversity. Furthermore, the research area shows a high frequency of landslides, presumably due to the instability of the water-soaked soils and the steepness of the slopes. It has been hypothesized that the weight of the mature forest triggers the incidence of these landslides (Bussmann et al. 2008). Where a landslide has occurred, the topsoil containing the organic layers and the humic horizon is removed and the deeper soil horizons or even the bedrock become exposed. Since these are very poor in nutrients, plant re-growth starting with cryp-

togams, a few orchids and ericaceous species, is usually very slow (Ohl & Bussmann 2004, Hartig & Beck 2003). Nevertheless, landslides apparently contribute
significantly to the dynamics of the vegetation of the RBSF and thus may enhance
plant diversity even though representing an element of landscape instability. This
observation is in agreement with the "intermediate disturbance hypothesis" (Levin
& Paine 1974, Connell 1978, Huston 1994), which states that medium disturbances
increase biodiversity. The impact of vegetation dynamics on plant diversity has
been well exemplified for the RBSF (chapters 4 and 7 of this volume). In addition,
recent studies by Hilt & Fiedler (2006) have shown that the species patterns of
moths differ substantially between areas of undisturbed and disturbed forests, the
latter being more diverse. The diversity of herbivorous insects apparently is likewise increased by vegetation dynamics.

Factors contributing to the maintenance of high biodiversity

Limitation of resources, especially of macronutrients prevents the dominance of
particular species which by excessive biomass production could outcompete others. This interrelation is known as the "soil nutrient hypothesis" of biodiversity
(Kapos et al. 1990, Huston 1994, Woodward 1996). The C:N - and the C:P – ratios of leaves increases with elevation indicating substantial N- and P-deficiencies
as compared to other forest types, especially the tropical lowland forest (Table 2).

Table 2. N- and P-content in leaves of trees (n = 5 ± SE) in the RBSF compared with data
(a) from McGroddy et al. (2004). Different upper case letters indicate statistically significant
differences; n = number of studies. From Soethe et al. (2006).

location/forest	C:N ratio	C:P ratio
RBSF (1900 m)	23.3 ± 1.2[a]	232 ± 21[a]
RBSF (2400 m)	38.2 ± 3.8[b]	430 ± 44[b]
RBSF (3000 m)	44.7 ± 1.1[b]	372 ± 19[b]
temperate broadleaved forest (a)	30.4 (n = 29)	357 (n = 28)
tropical lowland forest (a)	30.1 (n = 7)	951 (n = 12)

Wilcke et al (2008) showed an increase of soil organic matter, carbon stocks and
the C/N-ratio with elevation in the RBSF, while mineralization was negatively
correlated with altitude. Even the organic soil layer is quite poor in macronutrients.
As a likely consequence, woody plants at higher elevations invest more of their
biomass into the root system (Leuschner et al. 2007; chapter 8 of this volume). In
addition, the establishment of an effective mycorrhiza is of utmost importance for
the perennial plants. Very high colonization rates of arbuscular mycorrhiza were

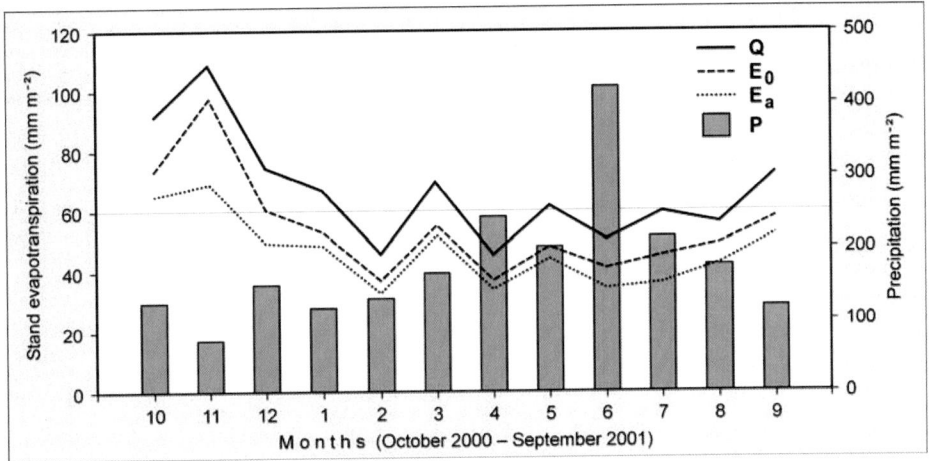

Figure 7. Stand parameters of the natural lower montane forest of the RBSF. Monthly sums of evapotranspiration, precipitation *(P)* in mm and monthly net radiation balance *(Q)*, here also expressed in mm as equivalent to vaporization. E_a= monthly actual, E_0= monthly potential evapotranspiration. During the wet months, E_a almost equals E_0. Low evapotranspiration rates contribute to low nutrient acquisition, in particular of mobile nutrients like nitrogen. From Motzer et al. (2007).

found not only in trees at all altitudes but also in hemi-epiphytic ericads, epiphytic orchids, ferns, and even in some groups of liverworts (Kottke et al. 2008; chapter 5 of this volume). The density of the litter-decomposing soil fauna, especially of the primary decomposers in the RBSF, is generally low and declines with altitude. In addition to low mineralization rates, nutrient uptake is also low. Because of the high degree of cloudiness together with a high relative humidity, evapotranspiration and, in turn, water and nutrient uptake rates are low (Fig. 7). During extremely wet months tree growth is even suspended. However, due to the elevational increase in precipitation leaching of nutrients increases with altitude, adding to the deterioration of the litter quality of the upper montane forest.

Table 3. Leaf area index as an indicator of the abiotic resource light in various habitats of the RBSF. Competition for light is especially strong in the lower montane forest. From Motzer et al. (2007) and Oesker (unpubl.).

site	gully	gully →slope	slope	slope	elfin forest
altitude [m]	2050	1950	1975	2125	2240
LAI [m²/m²]	9.8	6.7	6.4	5.9	3.0

Reduced light intensities (Table 3) in the dense forest, on the other hand, favour the occurrence of plant life forms such as lianas and epiphytes, which considerably contribute to plant diversity of the area. Favoured by the perhumid climate the wealth of epiphytic species is staggering: Up to 98 vascular epiphyte species have been counted on one single tree – the highest number ever recorded - and 225 on six trees (Werner et al. 2005). A very important result of the research reported here is the fact that keystone species of plants or animals could not be identified in the natural forest. Rather, a variety of up to 160 plant species share one hectare with only very few individuals of any one species present (Martinez 2006). The scattered populations of discrete tree species in the dense forest pose the question of the specificity of the biotic interactors, e.g. pollinators, seed dispersers and mycorrhizal fungi. *Vice versa* the outstanding diversity of herbivorous insects, e.g. moths, should also be reflected by a similar diversity of fodder plants. Although several specific interactions have already been detected, e.g. of bats or humming birds with particular flowers (Wolff 2006), biotic interaction is still an open field for further investigation. It should be borne in mind, however, that it is the multitude of rather weak biotic interactions which stabilizes the outstanding biodiversity of the ecosystem.

The influence of biodiversity on ecosystem processes and the question of redundancy

Ecosystems channel and circulate matter and energy, and exchange these with their surroundings. As most ecosystem processes are accomplished via food webs, the question arises as to the influence of the degree of diversity and food-web complexity on element cycling and matter balance.

With regard to an ecosystem we may differentiate two categories of elements, essential and luxury (redundant) components. In his security hypothesis, Schulze (1989; see also Schulze et al. 1999) compared these components and their functions with the parts of a car and their intended functions - upon normal use - and in emergency situations. A luxury part may facilitate normal driving, e.g. a brake booster, but will become a lifesaver in an emergency situation. For an ecosystem an emergency situation may result from a change in its abiotic setting or from a biological calamity. In the case of a biodiversity hotspot the question arises which species are necessary (and sufficient) for the "normal" functioning of the ecosystem: Which species may function in alternative processes ("exchangeability" of species), which as emergency components, and which may be redundant and dispensable (Gitay et al. 1996). In a more general sense the question is posed as to the resilience of a hotspot of biodiversity to unpredictable changes or events.

Ecosystems are commonly described by the flux of matter, including water, and of energy. The balance of these fluxes can be taken as a measure of the stability of the ecosystem. Characterization of the ecosystem, however, is by its biotic components and their interactions, i.e. of individuals in populations. These interac-

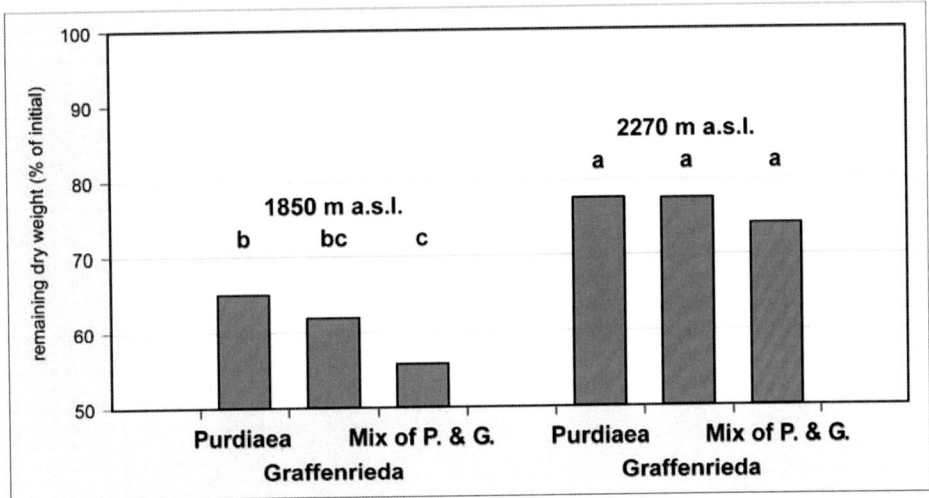

Figure 8. Effect of litter quality (C/N-ratio) and elevation on litter decomposition (measured as remaining dry weight). Leaf litter of *Graffenrieda emarginata* (C/N 42), *Purdiaea nutans* (C/N 74), and a mixture of both (mix) was exposed in the field for 12 month at two different altitudes. Bars sharing the same letter are not significantly different (Tukey's HSD test; P > 0.05). From Maraun et al. (2008), slightly modified.

tions may be strong in an ecosystem with a few keystone species (few degrees of freedom) but weaker and more variable in species-rich ecosystems (many degrees of freedom).

A nectar-licking bat (3 of at least 21 bat species in the RBSF belong to this guild, see Matt et al. 2008) must change its diet several times in the course of the year, as most perennial plants (e.g. *Macrocarpaea* and some Marcgraviaceae) show a pronounced seasonality of flowering. A high phytodiversity favours weak plant-animal interactions, i.e. counteracts a too tight specialization, thus reducing the risk of missing a special partner in a low density population. Recent work on host-plant relationships showed that in spite of, or even because of the narrower tropical habitat ranges, herbivores in tropical lowland forests are not generally more specialized than in temperate-zone ones, supporting the idea of preponderance of weak and often asymmetric biological interactions (Novotny et al. 2006). As already mentioned, keystone species with a wide habitat range could not be identified in the hotspot ecosystem represented by the RBSF. But the general attenuation of species diversity with increasing elevation poses the question whether the ecological functions and ecosystem services provided by many lower-montane species are taken over by a smaller number of less specific upper-montane species, or whether the lower montane and the upper montane forests represent different variants of the ecosystem (Fig. 9). Irrespective of the extreme cases, a gradual change of the ecosystem with increasing altitude appears to reflect the situation

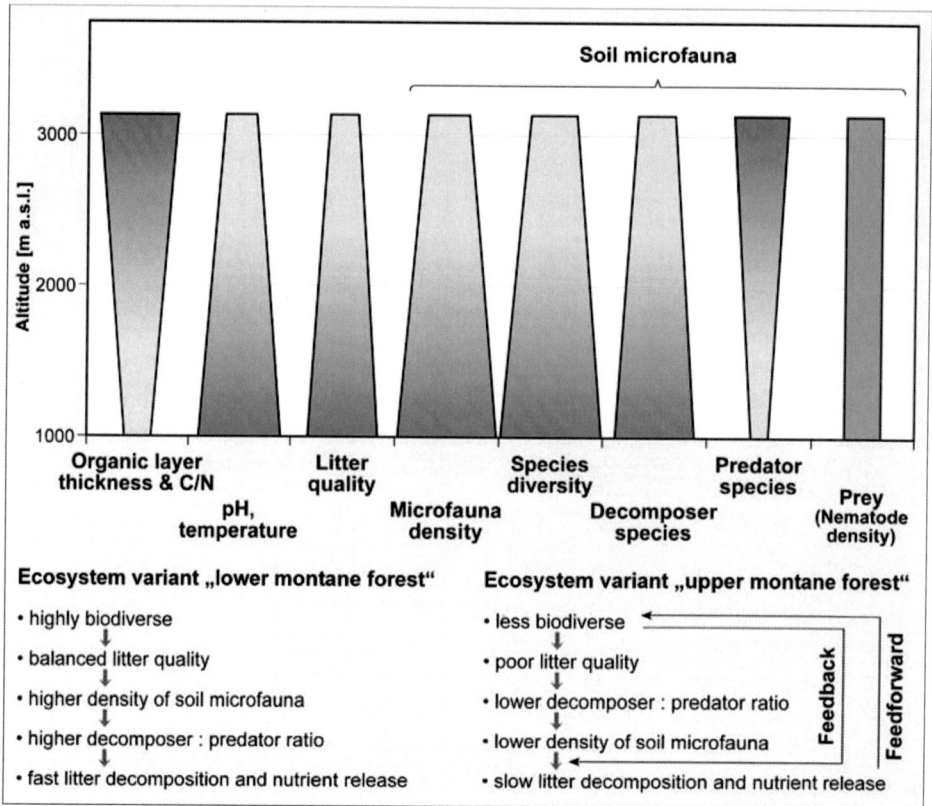

Figure 9. A possible feedback of biodiversity on ecosystem processes in lower montane (1850 m) and upper montane (2270 m) forest. A faster flux of resources in the lower montane forest promotes plant biodiversity, which in turn produces a more balanced litter quality.

best, concomitantly with a gradient of indispensability for individual components and their interactions.

Wilcke et al. (2008) and Iost et al. (2008) showed an accumulation of organic soil material at higher altitudes, which cannot only be attributed to a temperature-triggered decrease of the mineralization rate but is also caused by poorer litter "quality", i.e. lower nutrient contents (Table 2, Fig. 8). This concurs with an elevational change in the guild compositions of the soil fauna but also with a decrease of the population densities of litter decomposers. Comparison of the soil organic layer and carbon stocks of the lower and the upper mountain forest ecosystem of the RBSF reveal an effect of biodiversity on the respective matter fluxes. Due to a higher litter quality, the higher biodiversity encountered in the lower montane forest accomplishes a faster flux of the resources (nutrients) through the ecosystem, which in turn favours diversity of plant species (Homeier et al. 2008). In this case, a feedback of biodiversity on the ecosystem processes can be inferred (Fig. 9). Conceivably, such an effect could stabilize the tropical lower montane rainforest ecosystem, but we still lack empirical data. The fast recovery of a highly diverse secondary forest in the RBSF (Martinez et al. 2008) inhabited by an extraordinary diversity of herbivorous insects (Hilt & Fiedler 2008) nevertheless shows a substantial resilience of that ecosystem against small-scale disturbances, as long as recolonization by organisms lost through disturbance is possible.

The lower and upper montane forests demonstrate the existence of such a feedback, but we are not yet able to nail this effect down to the species level. This situation is best characterized by the "rivet"-hypothesis (Ehrlich 1994), which assumes that removal of a number of rivets (guild components) rather than of a specific rivet (a key species) will destabilize an airplane's wing (an ecosystem) to the point where it breaks. As a consequence, we cannot yet state whether or not there is species redundancy in our biodiversity hotspot.

Acknowledgements. The work of the research group: "Functionality in a Tropical Mountain Rainforest: Diversity, Dynamic Processes and Utilization Potentials under Ecosystem Perspectives" was sponsored by the German Research Foundation DFG (FOR 402) and by the Foundation "Nature and Culture International", San Diego.

References

Barthlott W, Mutke J, Rafiqpoor MD, Kier G, Kreft H (2005) Global centres of vascular plant diversity. Nova Acta Leopoldina NF 92, 342: 61-83

Beck E, Bendix J, Kottke I, Makeschin F, Mosandl R (eds) (2008a) Gradients in a Tropical Mountain Ecosystem of Ecuador. Ecological Studies 198. Springer, Berlin, Heidelberg, New York

Beck E, Makeschin F, Haubrich M, Richter M, Bendix J, Valerezo C (2008b) The Ecosystem (Reserva Biológica San Francisco). In: Beck E, Bendix J, Kottke I, Makeschin F, Mosandl R (eds) Gradients in a tropical mountain ecosystem of Ecuador. Ecological Studies 198. Springer, Berlin, Heidelberg, New York, pp 63-74

Bendix J, Homeier J, Cueva Ortiz E, Emck P, Breckle SW, Richter M, Beck E (2006) Seasonality of weather and tree phenology in a tropical evergreen mountain rain forest. International Journal of Biometeorology 50: 370-384

Bendix J, Rollenbeck R, Richter M, Emck P (2008) Climate. In: Beck E, Bendix J, Kottke I, Makeschin F, Mosandl R (eds) Gradients in a tropical mountain ecosystem of Ecuador. Ecological Studies 198. Springer, Berlin, Heidelberg, New York, pp 63-74

Braun H (2008) Tettigoniidae. In: Liede-Schumann S, Breckle S-W (eds) Provisional Checklist of Flora and Fauna of the San Francisco valley and its surroundings (Reserva

Biológica San Francisco, Province Zamora-Chinchipe, southern Ecuador). Ecotropical Monographs 4: 215-220

Brehm G, Pitkin LM, Hilt N, Fiedler K (2005) Montane Andean rain forests are a global diversity hotspot of geometrid moths. Journal of Biogeography 32: 1621-1627

Brehm G, Süssenbach D, Fiedler K (2003b) Unique elevational diversity patterns of geometrid moths in an Andean montane rainforest. Ecography 26:356-366

Bussmann RW, Wilcke W, Richter M (2008) Landslides as important disturbance regimes – causes and regeneration. In: Beck E, Bendix J, Kottke I, Makeschin F, Mosandl R (eds) Gradients in a tropical mountain ecosystem of Ecuador. Ecological Studies 198. Springer, Berlin, Heidelberg, New York, pp 325-338

Connell JH (1978) Diversity in tropical rainforests and coral reefs. Science 199:1302-1310

Cueva Ortiz E, Homeier J, Breckle SW, Bendix J, Emck P, Richter M, Beck E (2006) Seasonality in an evergreen tropical mountain rainforest in southern Ecuador. Ecotropica 12: 69-85

Dillon MO, Sagástegui AA, Sánchez VI, Llatas S, Hensold N (1995) Floristic inventory and biogeographic analysis of montane forests in northwestern Peru. In: Churchill SP, Balslev H, Forero E, Luteyn JL (eds) Biodiversity and Conservation of Neotropical Montane Forests. New York Botanical Garden, Bronx, pp 251-269

Ehrlich PR (1994) Energy use and biodiversity loss. Philosophical Transactions of the Royal Society of London 344: 99-104

Fiedler K, Brehm G, Hilt N, Süssenbach D, Onore G, Bartsch D, Häuser CH (2008) Moths (Lepidoptera: Arctiidae, Geometridae, Hedylidae, Pyraloidea, Sphingidae, Uraniidae) In: Liede-Schumann S, Breckle S-W (eds) Provisional Checklist of Flora and Fauna of the San Francisco valley and its surroundings (Reserva Biológica San Francisco, Province Zamora-Chinchipe, southern Ecuador). Ecotropical Monographs 4: 155-214

Gentry AH, Dodson CH (1987) Diversity and biogeography of neotropical vascular epiphytes. Annals of the Missouri Botanical Garden 74: 205-233

Gitay H, Wilson JB, Lee WG (1996) Species Redundancy: A Redundant Concept? Journal of Ecology 84: 121-124

Gradstein SR, Bock C, Mandl N, Nöske N (2008) Bryophyta: Liverworts and Hornworts. In: Liede-Schumann S, Breckle S-W (eds) Provisional Checklist of Flora and Fauna of the San Francisco valley and its surroundings (Reserva Biológica San Francisco, Province Zamora-Chinchipe, southern Ecuador). Ecotropical Monographs 4: 69-87

Hartig K, Beck E (2003) The bracken fern (*Pteridium arachnoideum* (Kaulf.) Maxon) dilemma in the Andes of Southern Ecuador. Ecotropica 9: 3-13

Haug I, Lempe J, Homeier J, Weiß M, Setaro S, Oberwinkler F, Kottke I (2004) *Graffenrieda emarginata* (Melastomataceae) forms Mycorrhizas with Glomeromycota and with a member of *Hymenoscyphus ericae* aggr. in the organic soil of a neotropical mountain rain forest. Canadian Journal of Botany 82: 340-356

213

Häuser CL (2008) Butterflies and skippers. In: Liede-Schumann S, Breckle S-W (eds) Provisional Checklist of Flora and Fauna of the San Francisco valley and its surroundings (Reserva Biológica San Francisco, Province Zamora-Chinchipe, southern Ecuador). Ecotropical Monographs 4: 145-154

Herzog SK, Kessler M, Bach K (2005) The elevational gradient in Andean bird species richness at the local scale: a foothill peak and a high-elevation plateau. Ecography 28: 209–222

Hilt N (2005) Diversity and species composition of two different moth families (Lepidoptera: Arctiidae vs. Geometridae) along a successional gradient in the Ecuadorian Andes. PhD Thesis, University of Bayreuth, Bayreuth URL: http://opus.ub.uni-bayreuth.de/volltexte/2006/201/

Hilt N, Brehm G, Fiedler K (2006) Diversity and ensemble composition of geometrid moths along a successional gradient in the Ecuadorian Andes. Journal of Tropical Ecology 22: 155 – 166

Hilt N, Fiedler K (2008) Successional stages of faunal regeneration. In: Beck E, Bendix J, Kottke I, Makeschin F, Mosandl R (eds) Gradients in a tropical mountain ecosystem of Ecuador. Ecological Studies 198. Springer, Berlin, Heidelberg, New York, pp 453-461

Holmgren M, Scheffer M, Excurra E, Gutiérrez JR, Mohren GMJ (2001) El Niño effects on the dynamics of terrestrial ecosystems. Trends in Ecology and Evolution 16: 89-94

Homeier J, Werner FA (2008) Spermatophyta. In: Liede-Schumann S, Breckle S-W (eds) Provisional Checklist of Flora and Fauna of the San Francisco valley and its surroundings (Reserva Biológica San Francisco, Province Zamora-Chinchipe, southern Ecuador). Ecotropical Monographs 4: 15-58

Homeier J, Werner FA, Gradstein SR, Breckle S-W, Richter M (2008) Potential vegetation and floristic composition of Andean forests in South Ecuador. In: Beck E, Bendix J, Kottke I, Makeschin F, Mosandl R (eds) Gradients in a tropical mountain ecosystem of Ecuador. Ecological Studies 198. Springer, Berlin, Heidelberg, New York, pp 87-102

Huston MA (1994) Biological diversity – the coexistence of species in changing landscapes. Cambridge University Press, Cambridge

Illig J, Langel R, Norton RA, Scheu S, Maraun M (2005) Where are the decomposers? Uncovering the soil food web of a tropical montane rain forest in southern Ecuador using stable isotopes (15N). Journal of Tropical Ecology 21: 589–593

Illig J, Maraun M (2008) Oribatid mites from the Reserva San Francisco, Ecuador. In: Liede-Schumann S, Breckle S-W (eds) Provisional Checklist of Flora and Fauna of the San Francisco valley and its surroundings (Reserva Biológica San Francisco, Province Zamora-Chinchipe, southern Ecuador). Ecotropical Monographs 4: 221-230

Iost S, Makeschin F, Abiy M, Haubrich F (2008) Biotic soil activities. In: Beck E, Bendix J, Kottke I, Makeschin F, Mosandl R (eds) Gradients in a tropical mountain ecosystem of Ecuador. Ecological Studies 198. Springer, Berlin, Heidelberg, New York, pp 221-232

Jørgensen PM, Leon-Yanez S (1999) Catalogue of the vascular plants of Ecuador. Monographs in Systematic Botany from the Missouri Botanical Garden 75

Jørgensen PM, Ulloa Ulloa C (1994) Seed plants of the high Andes of Ecuador – a checklist. Aarhus University (AAU) Reports 34

Jørgensen PM, Ulloa Ulloa C, Madsen JE, Valencia R (1995) A floristic analysis of the high Andes of Ecuador. In: Churchill SP, Balslev H, Forero E, Luteyn JL (eds) Biodiversity and Conservation of Neotropical Montane Forests. New York Botanical Garden, Bronx, pp 21-237

Kapos V, Pallant E, Bien A, Freskos S (1990) Gap frequencies in lowland rainforest sites on contrasting soils in Amazonian Ecuador. Biotropica 22: 218-225

Kier G, Mutke J, Dinerstein E, Ricketts TH, Küper W, Kreft H, Barthlott W (2005) Global patterns of plant diversity and floristic knowledge. Journal of Biogeography 32: 1107-1116

Kleidon A, Mooney HA (2000) A global distribution of biodiversity inferred from climatic constraints: results from a process-based modelling study. Global Change Biology 6: 507-523

Kottke I, Beck A, Haug I, Setaro S, Jeske V, Suarez JP, Pamino L, Preußig M, Nebel M, Oberwinkler F (2008) Mycorrhizal state and new and special features of mycorrhizae of trees, ericads, orchids, ferns, and liverworts. In: Beck E, Bendix J, Kottke I, Makeschin F, Mosandl R (eds) Gradients in a tropical mountain ecosystem of Ecuador. Ecological Studies 198. Springer, Berlin, Heidelberg, New York, pp 139-150

Krashevskay V (2008) Testacea. In: Liede-Schumann S, Breckle S-W (eds) Provisional Checklist of Flora and Fauna of the San Francisco valley and its surroundings (Reserva Biológica San Francisco, Province Zamora-Chinchipe, southern Ecuador). Ecotropical Monographs 4: 231-236

Krömer T, Kessler M, Gradstein SR, Acebey A (2005) Diversity patterns of vascular epiphytes along an elevational gradient in the Andes. Journal of Biogeography 32: 1799-1809

Kürschner H, Parolly G (2008) Bryophyta: Mosses. In: Liede-Schumann S, Breckle S-W (eds) Provisional Checklist of Flora and Fauna of the San Francisco valley and its surroundings (Reserva Biológica San Francisco, Province Zamora-Chinchipe, southern Ecuador). Ecotropical Monographs 4: 89-100

Lehnert M, Kessler M, Salazar LI, Navarrete H, Werner F, Gradstein SR (2008) Pteridophytes. In: Liede-Schumann S, Breckle S-W (eds) Provisional Checklist of Flora and Fauna of the San Francisco valley and its surroundings (Reserva Biológica San Francisco, Province Zamora-Chinchipe, southern Ecuador). Ecotropical Monographs 4: 59-68

Leuschner C, Moser G, Bertsch C, Röderstein M, Hertel D (2007) Large altitudinal increase in tree root/shoot ratio in tropical mountain forests of Ecuador. Basic and Applied Ecology 8: 219-230

Levin SA, Paine RT (1974) Disturbance, patch formation, and community structure. Proceedings of the National Academy of Sciences USA 71: 2744-2747

Liede-Schumann S, Breckle S-W (2008) Provisional Checklist of Flora and Fauna of the San Francisco valley and its surroundings (Reserva Biológica San Francisco, Province Zamora-Chinchipe, southern Ecuador). Ecotropical Monographs 4

Litherland M, Aspden JA, Jemielita RA (1994) The metamorphic belts of Ecuador. Overseas Memoir of the British Geological Survey 11

Maraun M, Illig J, Sandman D, Krashevskaya V, Noton RA, Scheu S (2008) Soil Fauna. In: Beck E, Bendix J, Kottke I, Makeschin F, Mosandl R (eds) Gradients in a tropical mountain ecosystem of Ecuador. Ecological Studies 198. Springer, Berlin, Heidelberg, New York, pp 185-196

Martinez J JA, Mahecha MD, Lischeid G, Beck E (2008) Succession stages of vegetation regeneration: Secondary tropical mountain forests. In: Beck E, Bendix J, Kottke I, Makeschin F, Mosandl R (eds) Gradients in a tropical mountain ecosystem of Ecuador. Ecological Studies 198. Springer, Berlin, Heidelberg, New York, pp 419-426

Martínez Jerves JA 2006 Los Bosques Secundarios en el Sur del Ecuador: Análisis de bosques secundarios montanos lluviosos relevan diferentes rutas de regeneración de bosque, dependiendo del tipo de impacto. PhD Thesis, University of Bayreuth

Matt F (2001) Pflanzenbesuchende Fledermäuse im tropischen Bergregenwald: Diversität, Einnischung und Gildenstruktur – Eine Untersuchung der Fledermausgemeinschaften in drei Höhenstufen der Andenostabdachung des Podocarpus Nationalparks in Südecuador. PhD Thesis, University of Erlangen-Nürnberg

Matt F, Almeida K, Arguero A, Reudenbach C (2008) Seed disperal by birds, bats, and wind. In: Beck E, Bendix J, Kottke I, Makeschin F, Mosandl R (eds) Gradients in a tropical mountain ecosystem of Ecuador. Ecological Studies 198. Springer, Berlin, Heidelberg, New York, pp 161-170

McGroddy ME, Daufresne T, Hedin LO (2004) Scaling of C:N:P stoichiometry in forests worldwide: implications of terrestrial redfield-type ratios. Ecology 85: 2390-2401

Morrison JC, Olson DM, Loucks CJ, Dinerstein E, Allnutt TF, Wikramanayake ED, Ricketts TH, Burgess ND, Kura Y, Powell GVN, Lamoreux JF, Underwood EC, Wettengel WW, D'Amico JA, Hedao P, Itoua I, Kassem KR, Strand HE (2001) Terrestrial Ecoregions of the World. A New Map of Life on Earth. BioScience 51: 933-938

Moser G, Hertel D, Leuschner C (2007) Altitudinal change in LAI and stand leaf biomass in tropical montane forests: a transect study in Ecuador and a pan-tropical meta-analysis. Ecosystems 10: 924-935

Motzer T, Munz N, Küppers M, Schmitt D, Anhuf D (2005) Stomatal conductance, transpiration and sap flow of tropical montane rain forest trees in the southern Ecuadorian Andes. Tree Physiology 25: 1283-1293

Myers N, Mittermaier RA, Mittermaier CG, Fonseca GAB da, Kent J (2000) Biodiversity hotspots for conservation priorities. Nature 403: 853-858

Niemann H, Behling H (2008) Late Quaternary vegetation, climate and fire dynamics inferred from the El Tiro record in the southeastern Ecuadorian Andes. Journal of Quaternary Science: DOI 10.1002/jqs.1134

Nöske N, Sipman HJ (2008) Lichens. In: Liede-Schumann S, Breckle S-W (eds) Provisional Checklist of Flora and Fauna of the San Francisco valley and its surroundings (Reserva Biológica San Francisco, Province Zamora-Chinchipe, southern Ecuador). Ecotropical Monographs 4: 101-117

Novotny V, Drozd P, Miller SE, Kulfan M, Janda M, Basset Y, Weiblen GD (2006) Why are there so many species of herbivorous insects in tropical rainforests? Science 313: 1115-1118

Ohl C, Bussmann RW (2004). Recolonisation of natural landslides in tropical mountain forests of Southern Ecuador. Feddes Repertorium 115: 248-264

Paulsch D (2008) Aves. In: Liede-Schumann S, Breckle S-W (eds) Provisional Checklist of Flora and Fauna of the San Francisco valley and its surroundings (Reserva Biológica San Francisco, Province Zamora-Chinchipe, southern Ecuador). Ecotropical Monographs 4: 131-144

Rahbek C (2005) The role of spatial scale and the perception of large-scale species-richness patterns. Ecology Letters 8: 224-239

Rasmussen JF, Rahbek C, Horstman E, Poulsen MK, Bloch H (1994) Aves del Parque Nacional Podocarpus, una lista anotada. CECIA, Quito, Ecuador

Richter M (2001) Vegetationszonen der Erde. Klett-Perthes, Gotha, Stuttgart

Richter M, Diertl KH, Peters T, Bussmann RW (2008) Vegetation structures and ecological features of the upper timberline ecotone. In: Beck E, Bendix J, Kottke I, Makeschin F, Mosandl R (eds) Gradients in a tropical mountain ecosystem of Ecuador. Ecological Studies 198. Springer, Berlin, Heidelberg, New York, pp 125-138

Sagástegui AA, Dillon MO, Sánchez Vega I, Leiva Gonzales S, Lezama Asencio P (1999) Diversidad Florística del Norte de Perú, Vol. I. WWW Peru Programme Office and Fondo Editorial, Trujillo

Sagástegui AA, Sánchez Vega I, Zapata Cruz M, Dillon MO (2004) Diversidad Florística del Norte de Perú, Vol. 2. Bosques montanos. WWW Peru Programme Office and Fondo Editorial, Trujillo

Schulze E-D (1989) Ökosystemforschung-die Entwicklung einer jungen Wissenschaft. In: Gerwin R (ed) Wie die Zukunft Wurzeln schlug. Springer, Berlin Heidelberg New York, pp 55-64

Schulze E-D, Mooney HA (1993) Ecosystem function of biodiversity: a summary. In: Schulze E-D, Mooney HA (eds) Biodiversity and Ecosystem Function. Ecological Studies 99. Springer, Heidelberg, New York, pp 497-510

Soethe N, Lehmann J, Engels C (2006) The vertical pattern of rooting and nutrient uptake at different altitudes of a south Ecuadorian montane forest. Plant Soil 286(1-2): 287-299

Stevens GC (1989) The latitudinal gradient in geographic range: how so many species coexist in the tropics. American Naturalist 133: 240-256

Stevens GC (1992) The elevational gradient in altitudinal range: an extension of Rapoport's latitudinal rule to altitude. American Naturalist 140: 893-911

Stevens GC (1996) 1996. Extending Rapoport's rule to Pacific marine fishes. Journal of Biogeography 23: 149–154

Weigend M (2002) Observations on the Biogeography of the Amotape-Huancabamba Zone in Northern Peru. Botanical Review 68: 38-54

Weigend M (2004) Additional Observations on the Biogeography of the Amotape-Huancabamba Zone in Northern Peru: Defining the Southeastern Limits. Revista Peruana de Biología 11: 127-134

Werner FA, Homeier J, Gradstein SR (2005) Diversity of vascular epiphytes on isolated remnant trees in the montane forest belt of southern Ecuador. Ecotropica 11: 21-40

Wilcke W, Yasin S, Schmitt A, Valerezo C, Zech W (2008) Soils along the altitudinal transect and in catchments. In: Beck E, Bendix J, Kottke I, Makeschin F, Mosandl R (eds) Gradients in a tropical mountain ecosystem of Ecuador. Ecological Studies 198. Springer, Berlin, Heidelberg, New York, pp 75-86

Wolff D (2006) Nectar sugar composition and volumes of 47 species of Gentianales from a southern Ecuadorian montane forest. Annals of Botany 97: 767-777

Woodward CL (1996) Soil compaction and topsoil removal effects on soil properties and seedling growth in Amazonian Ecuador. Forest Ecology and Management 82: 197-209